FutureWorld

Where science fiction becomes science

First published 2008 by Boxtree
an imprint of Pan Macmillan Ltd
Pan Macmillan, 20 New Wharf Road, London N1 9RR
Basingstoke and Oxford
Associated companies throughout the world
www.panmacmillan.com

ISBN 978 0 75222 672 9

9 8 7 6 5 4 3 2 1

A CIP catalogue record for this book is available from the British Library.

Edited by Damon McCollin-Moore
Picture research by Felicity Page
Design by Dan Newman/Perfect Bound Ltd
Printed in Great Britain by Butler and Tanner, Somerset

FutureWorld

Where science fiction becomes science

Professor Mark L. Brake & Reverend Neil Hook

BOXTREE

CONTENTS

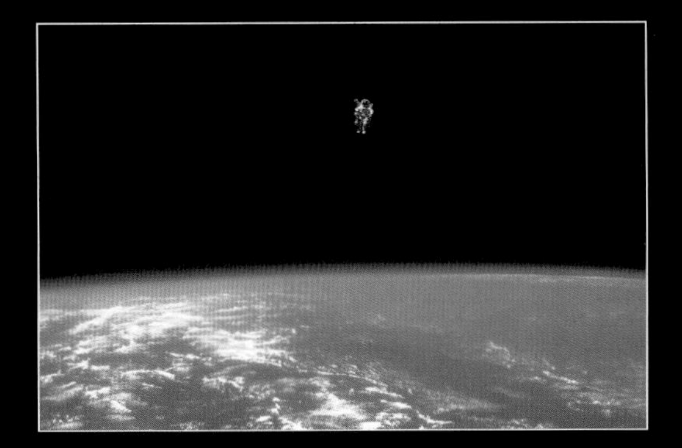

FOREWORD

Part of the tension between science and the public stems from the fact that good science is based on dispassionate observation, the impartial gathering of evidence and the unbiased testing of hypotheses. Whilst this method makes for robust science, it can make it difficult for non-scientists to grapple with.

Scientists have necessarily raised objectivity and the establishment of scientific 'truths' beyond the reach and influence of belief, ambition and prejudice. And when the science – as the best science should be – is completely isolated from these influences and emotions, and out of *our* reach, the scientists ask us to trust them.

Science needs our trust but it is in a bind. Unlike many other occupations that play a vital role in society, scientists cannot make a virtue of their subjectivity and personal beliefs as a reason to support what they do; they cannot kiss a baby or tell a sad childhood story in support of their research. If scientists wish to remain credible as practitioners, they have to 'let the science do the talking'. And, for the average person, the science is hard to understand.

We want to know what science entails, how it's made and what its objectives are – how else are we to decide if we want it? And scientists want to communicate their work and its significance and benefits – how else are they to gain our support?

Good science fiction has always acted as a conduit between science and the public, an instantaneous translator passing back and forth information about theories, breakthroughs, anxieties and vetoes.

FutureWorld is about the inspiration that science and fiction constantly give to one another. Each of its one hundred cross-referenced entries is a celebration of this mutually reinforcing relationship. Ranging from the monumental (the atomic bomb and black holes) to the mundane (screensavers and automated vacuum cleaners), this book charts the ongoing dialogue between science and the public about which direction to take as we plot our course into the future.

Note: words highlighted **LIKE THIS** in the text refer to the titles of other entries.

INTRODUCTION

This book is about the future. The future as it was imagined in the fiction of the past. The future we now inhabit.

Today, media headlines trumpet the discovery of extra solar planets, cloning experiments and the teleportation of atoms. On the television we see walking, talking robots, private jets that ferry travellers to the edge of space, and interplanetary probes that rendezvous with asteroids. Some scientists claim that the first human to live for a thousand years has already been born. We live in a science fictional world.

The food we eat, the clothes we wear, the culture we consume, the pre-natal care our mothers took – all these things shape our physical and mental development. All of these things were once the stuff of fiction that has now become fact. The days when only fictional scientists performed unimaginable procedures like transplanting human hearts are long gone.

But there are downsides. A crumbling environment, nuclear stockpiles, continuous surveillance, rogue pathogens and schoolchildren on psychotropic medication to name but a few. On such a threatened planet, science fiction has become hard realism. It seems somehow laced into global atrocities, such as 9/11, and imminent threats, such as avian flu.

We need science fiction more than ever. It can act as a vital forum for thinking through the shape of things to come. A dizzying display of possible futures beguiles and torments us. There are thousands of challenges that the human race must face to accommodate our burgeoning populations, feed our starving children, and ease our troubled world.

We no longer think of science fiction as a sub-culture. Gone are the days of the pulp fiction magazines of the 1920s and 30s. Today's proliferation of science fiction in the form of books, films, television, games and comics reflects its increasing impact. Science fiction has moved into the mainstream with the advent of the information age it helped realise – the digital and hyperconnected globe of the World Wide Web. At the time of writing, science fiction accounted for twenty-two of the top fifty grossing movies of all time. Audiences of all ages will pay a tenner each to watch the latest science fiction blockbuster on the big screen. Eight million viewers regularly tune in to the BBC to watch *Doctor Who*. Science fiction even has its own channels running back-to-back series and films. And in the ever expanding field of computer gaming, science fiction titles dominate. The fastest-selling media product in history, according to BBC Online, was Microsoft's science fiction video-game *Halo 3*, with sales of the game generating US$170 million on its first day!

So, what *is* science fiction?

It's a good question and one that many writers and scientists have tried to answer. American author Robert Heinlein felt that a good start is to say that science fiction is realistic speculation about the future, based solidly on an understanding of nature and science.

And even though science fiction is a kind of fantasy, it's nonetheless different to the likes of Tolkien's *Lord of the Rings* and Philip Pullman's *His Dark Materials*. Those types of books are knowingly magical and fantastic

The difference with science fiction is this. It denies it's fantastic. You can tell something is science fiction by the way it uses an atmosphere of science to ease the willing suspension of disbelief that we normally associate with fantasy. That is the way science fiction makes imaginative speculations about the future.

So, we can think of science fiction as a controlled way to think and dream about the future. Science fiction conjures an infinity of nightmares and visions. And these visions have in common an interconnected mood and attitude, which is one part conscious science and one part unconscious dreamscape.

Science fiction is the literature of change. It's that class of fiction that concerns itself with the shifting currents in science and society. It's not conservative, in the way of old style fantasy. Science fiction tends to be radical. It thinks about the direction in which science is taking us. It criticises, extrapolates and revises. It asks questions of science in society, such as 'what if?' and 'if this goes on…'

When did science fiction begin?

It emerged along with science. Way back at the time of the Scientific Revolution, the world became an alien planet. When Copernicus made the Earth-shattering suggestion that we did not live at the centre of the Universe, his revolution cut two ways. It made 'Earths' of the planets, and it also brought the alien to Earth. The universe of our ancestors had been small, static and Earth-centred. Humanity was its guiding light. The new Universe was decentralised, inhuman and dark.

The first science fiction, stories of space voyages, was a response to the shock created by Copernicus' discovery. And ever since, science fiction has been a kind of human project that tries to make sense of the non-human universe in which we find ourselves.

FutureWorld is divided into four conceptual themes: **SPACE**, **TIME**, **MACHINE** and **MONSTER**. These staples of the genre (just try to name a science fiction tale that does not have one of these themes at its heart!) allow an exploration of the relationship between us, the human, and the non-human aspects of the Universe that science has revealed to us.

In the 1997 movie of Carl Sagan's novel *Contact*, the most dramatic episode occurs when Jodie Foster's character, Dr Ellie Arroway, goes on a galactic space flight. She is confronted with the visual marvels to be seen at the centre of our Galaxy and the awe and wonder of the Universe. She is lost for words in the face of such beauty and humbly suggests a poet may have been a better choice for such a journey.

That's just how science fiction works. It sometimes takes a poet to best express the taste, the feel and the human meaning of scientific discoveries. It is an attempt to put the stamp of humanity back onto the Universe and to make human what is alien.

Science fiction has helped to shape the way we see and do things, the way we dream of things to come. It has helped us to discover the familiar in relation to the unfamiliar, the ordinary in relation to the extraordinary and has forced us to explore the nature and limits of our own reality.

It has helped us build this future we live in.

SPACE

It's the 'final frontier' in *Star Trek*. It's where 'no-one can hear you scream' in *Alien* (1979). And it's from where Wells' Martians in *The War of the Worlds* (1898) 'regarded this Earth with envious eyes'.

As we can see from these three famous examples, the science fiction of space focuses on some facet of the natural world.

Sometimes space suggests the outward urge that is associated with the mastery and conquest of vast interstellar depths, sought by Captain Kirk and the legions of fleet and nimble ships of space opera.

At other times, it is the vast, cold and unsympathetic theatre of space, featured in *Alien*, and invoking the immense vacancy that we may never come to terms with. The unfathomable darkness of space in much fiction reminds us that life is precious and frail, in a Universe that is largely inhuman and deserted.

Wells' Martians are agents of the void. They were the first 'menace from space'; a timely reminder that we may not be at the top of the Universe's evolutionary ladder.

Much of the science fiction of space can be understood as a longing to escape our sense of being merely human. Earth is our prison. That's why we get tales in which, often through the marvels of space travel, the wonders and potential terrors of the Universe are explored, bringing tales of contact with extraterrestrial beings.

The space theme shows us that there are great similarities between science fiction and science. Science fiction is an imaginative device for doing a kind of theoretical science: the exploration of imagined worlds.

Scientists build models of hypothetical worlds, and then test their theories. Albert Einstein was famous for this. His thought experiments – *Gedankenexperimenten* – led to his Special Relativity theory. The science fiction writer also explores hypothetical worlds but with more freedom. Whilst scientists are generally bound to work within science's laws, science fiction's boundaries are much less well defined. But we can see that a spirit of 'what if?' is common to both endeavours.

There are many examples where science fiction has proposed theories far too speculative for the science of the day, but which have later proved to be prophetic. The theme of space

contains some great examples – space travel, satellites, men on the Moon – to name just a few. But it's crucial to remember that with science fiction the correctness of the science is not as important as its poetry. The emphasis is on the sense of wonder and adventure experienced in pursuit of the science itself.

Through most of its history, science fiction has held a positive view of the possibility of alien life. Much of science fiction is made up of just two sciences: physics and biology. Historically, physics came a lot earlier. Copernicus shifted the centre of the Universe way back in 1543. Just over a century later, Newton had produced a 'system of the world', one of the first attempts to produce a theory of everything.

As a result, early fictional accounts of alien life, from Kepler's *Somnium* (1634) to Wells' *War of the Worlds* (1898), were based mostly on the physics in question. Simply put, the argument was based on the idea that since there are billions of stars in the Universe, there must be millions, if not billions, of planets – plenty of places for the extra-terrestrial to live.

The line of reasoning is much like the principle of plenitude: everything that can happen will happen, and in a Universe fit for life, many planets will bear bug-eyed monsters. After all, it makes a far better story when you have books to sell!

So, the mere physics of the matter has been fiction's main concern, biology didn't come into it. The influence of science fiction was fed back into science, and by the twentieth century, a generation of scientists were cast under the same spell, and huge investments were made in a serious search for the alien.

Meanwhile, a unified theory of biology didn't see the light of day until the second half of the twentieth century. It brought a modern synthesis of aspects of biology and was accepted by the great majority of working biologists. And this evolutionary synthesis provides a new perspective on the question of extra-terrestrial life. Biologists are now far more sceptical about the possibility of complex alien life, let alone intelligence.

We may, after all, be alone in the Universe …

MEN ON THE MOON

Blame Galileo.

After all, he wielded the newly-invented telescope like a weapon of discovery. A new universe was unveiled, and the Moon was key. Galileo's revolutionary pamphlet *The Starry Messenger* told the tale of his discoveries with the telescope. Written in 1610, in it he urged his readers to imagine walking on the lunar mountains and craters, just like on Earth.

It was the first time the Moon became a real object for the great majority of people. Before Galileo, the Moon was just a disc in the sky. True, it had been imagined in Greek fiction, and in the seventeenth century it was the developing fictional obsession of the age. But with the telescope it truly became an object of wonder, as we began to contemplate the possibilities: is there life there? May we one day walk over that craggy terrain?

Years before the publication of *The Starry Messenger*, Johannes Kepler and Francis Godwin had begun imagining journeys to our **SATELLITE** (indeed, the word 'satellite' was actually coined by Kepler). Kepler's book, *Somnium*, though published in 1634, had first germinated in his mind as early as 1593. It was one of the first fictional Moon voyages with a strong scientific flavour. In the book, Kepler imagined alien life fit for a lunar landscape.

Godwin's *The Man in the Moone* (1638) was equally extraordinary. Again, it first gelled in Godwin's imagination in the 1590s. It explored the possibility of a space voyage to another world. And get this: Godwin's is the first English book in history to portray alien contact. *The Man in the Moone* captured the imagi-

❸ Long imagined, men finally set foot on the Moon at 10.56pm (Eastern Daylight Time) on 20 July 1969.
❹ Godwin's *Man in the Moone* (1638); to the Moon by, er, goose...

nation of John Wilkins, First Secretary of the Royal Society. Wilkins' own work was revised to take account of the popularity of Godwin's work, and the notion that it was just a matter of time before a lunar encounter took place. Wilkins proposed a flying machine would one day wing its way moonwards.

The Moon holds huge significance for science fiction. It was with his lunar speculations that Kepler invented the genre. For with scientific discovery goes storytelling, a key aspect of the human experience. With *Somnium,* Kepler realised that to understand the Moon it was not enough to put Galileo's observations into words; the words themselves had to be transformed by a new sort of fiction.

Put another way, the 're-discovery' of the Moon crystallized the relationship of science with science fiction. Just as Galileo seemed to nail the Moon for science, discovering mountains, craters and an Earth-like terrain, at that very moment the reality of the Moon once more receded from us. It was an alien landscape. Even though a discovery had been made for science, even more questions were lit in the nervous system of Kepler. And he began to use creativity, imagination and fiction to try to answer some of those questions.

Cyrano de Bergerac followed suit in spectacular style. According to Arthur C. Clarke, Cyrano's *The States and Empires of the Moon* (1657) is to be credited for conceiving the ramjet, a form of jet engine that contains no moving parts. Cyrano dreamt up a clever lunar culture, far outstripping its earthly equivalent. But by the time the cosmic voyage was taken seriously in the mid-nineteenth century, scientifically, lunar life held no credibility – the Moon was dead.

And yet, at little over a light second away, it was there to be conquered and claimed for science. For pulp fiction writers in particular, reaching the Moon became an article of faith. Foremost was Robert A. Heinlein. Books such as his *Rocket Ship Galileo* (1947) portray our satellite as a stepping-stone for the development of the solar system at large. Crucially, Heinlein's *The Man Who Sold the Moon* (1950) told a tale of the fight to finance the first Moon-shot, and how to sell the myth of space conquest to the world.

Fact followed fiction. Destination Moon became an obsession for Cold War politicians who saw the propagandist coup that a manned Moon landing signified. Apollo got there first. And a dozen US astronauts from the various missions were the only humans to set foot on lunar soil. Allegedly...

10 scientists who wrote science fiction

- Isaac Asimov (1920–1992)
- Francis Bacon (1561–1626)
- Arthur C. Clarke (1917–2008)
- John G. Cramer (1934–)
- Fred Hoyle (1915–2001)
- Johannes Kepler (1571–1630)
- Geoffrey A. Landis (1955–)
- Carl Sagan (1934–1996)
- Leo Szilárd (1898–1964)
- Konstantin Tsiolovsky (1857–1935)

FORCE FIELDS

All matter is made of atoms. Atoms are bound by forces. Take away atoms. Leave the forces behind. That's a force field.

Well, at least that's a force field in science fiction. It's a concept we all know and love. It's what happens when bug-eyed monsters, rogue asteroids, or a newly severed limb of Darth Vader are winging our way. Just slap up a force field projector and pour a pan-galactic gargle blaster. Job done.

The first force field found its way into fiction with E.E. 'Doc' Smith's *Skylark* and *Lensman* books in the 1930s and 40s. A more modern version is the 'deflector' on the *Enterprise* in *Star Trek*.

The truth is more problematic. As yet there is no known gadget capable of repelling all objects and energies. But we're working on it.

Scientists at NASA's Kennedy Space Centre and the NASA Institute for Advanced Concepts are researching the possibility of electric shields for Moon bases. Most of the deadly radiation in space is made up of electrically charged particles. So why not use a powerful electric field that has the same charge as the incoming radiation, thus deflecting the radiation away?

Now *that's* a force field.

LOST WORLDS

Until science fiction came along, lost worlds were unheard of. Before the 1770s, large parts of the world remained unknown to Europeans. As science and technology developed, and as modern 'civilisation' crept around the globe, fantastic travellers' tales became very popular.

A typical tale would first find our adventurer somewhere in the civilised world, usually London. Armed with a tall story, or an ancient scroll, our hero sets off to unknown lands and lost civilisations to find secret powers of great antiquity. The influence of these classic lost world stories of the 1880s, such as H. Rider Haggard's *King Solomon's Mines* (1886) and *She* (1887), can still be seen in the *Indiana Jones* movies and in video games such as Lara Croft's adventures in the *Tomb Raider* series.

The quintessential lost world was Plato's *Atlantis*. It became an integral part of the Scientific Revolution when Francis Bacon wrote *New Atlantis* (1626), about a mythical but exemplary land where rational scientists run society.

In the rapidly shrinking world of the twentieth century, fewer parts of the planet were left untouched. There were no more lost 'arks' to raid. Writers found it ever more difficult

Jessica Alba as Susan Storm fends off unwanted attention with the use of her force field in *Fantastic Four* (2005).

A normal day in the life of a lost world – *Jurassic Park* (1993).

Sputnik One: the dawn of the space age.

to plonk their heroes in strange places that remained mysterious and unmapped. Heroes were shipped instead to other Earths. It took a dash of DNA chicanery to build a lost world back on Earth, with Michael Crichton and Steven Spielberg mutating the menacing dinosaurs in *Jurassic Park* (1993).

SATELLITES

It's strange to think that when you phone a friend in a foreign country or program your 'Satnav' the signal is probably bounced up into space and back again down to Earth. Even the houses we live in are adorned with dishes designed to receive satellite signals from space. Used for far more than calls to the USA or watching the World Cup, the satellite has revolutionised our lives.

The most familiar satellite to grace the night sky is the Moon, Earth's naturally occurring satellite and companion. It was natural then that the first mention of an artificial satellite should be referred to as a 'brick moon'. It is the eponymous tale authored by Edward Everett Hale in 1861. The adventure begins when the heroes are accidentally launched into space aboard this proto-navigational aid. It is another accident which results in the launch of a satellite in Jules Verne's *The Begum's Millions*. In this 1878 novel a projectile is shot from a massive cannon so powerfully that it enters the Earth's orbit. One of the protagonists writes in a letter 'we saw your perfect shell, at forty-five minutes and four seconds past eleven, pass above our town. It was flying towards the west, circulating in space, which it will continue to do until the end of time. A projectile, animated with

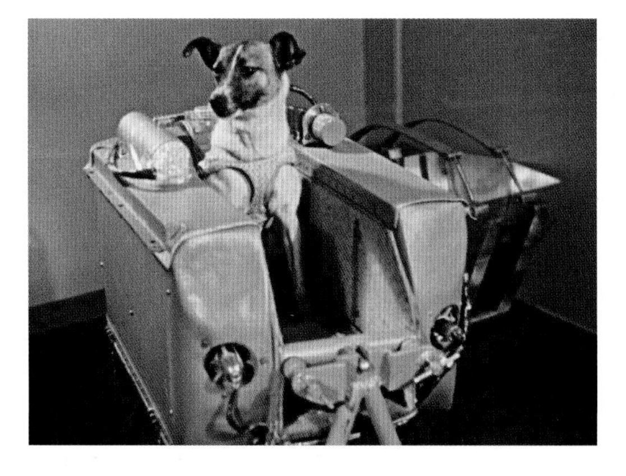

an initial speed twenty times superior to the actual speed, being ten thousand yards to the second, can never fall! This movement, combined with terrestrial attraction, destines it to revolve perpetually round our globe.'

As with both the **SPACE STATION** and the space elevator, it is the Russian science fiction author and scientist Konstantin Tsiolkovsky that we have to thank for some of the key theoretical developments concerning satellites. An avid reader of both Verne and Wells, it is likely that he would have had their stories in mind when he calculated the minimal orbital speed of satellites in a 1903 paper. In that paper Tsiolkovsky described the concept of geostationary satellites, which orbit the Earth over exactly the same spot, allowing a network to be built. It was, however, left to the German scientist Herman Potocnik to work out the actual geostationary distance (35,786 km) in his 1928 book *The Problem of Space Travel – The Rocket Motor.*

Science fiction's 'outside the box' style of thinking no doubt contributed to the development of the geostationary communications relay satellites which now encircle the globe. However, it was not through a story or film but rather an author whose innovative background led to him directly proposing the concept. Arthur C. Clarke famously outlined the model in a short technical paper in *Wireless World* magazine in 1954.

Although there was some preliminary work done on satellite design and launch capabilities it was not until 29 July 1955 that the USA announced its plan to place satellites in space. The self-imposed deadline was spring of 1958 for the completion of what came to be known as Project Vanguard. The USSR response was to announce that a Soviet satellite would be launched by the end of 1957. The space race had begun.

The Soviets remained true to their word and placed Sputnik One in orbit on 4 October 1957. This was quickly followed by Sputnik Two a month later, nicknamed 'Muttnik' because it featured both the first living creature launched in space (Laika the space dog) and the first space casualty (when she did not return safely). From that point on space has become increasingly cluttered by a mixture of different types of satellites. One recent estimate placed only 7% of the approximately nine thousand artificial objects orbiting the Earth from space as active. The others are debris of one sort or another, destined to either burn up on re-entry or continue their journey around the globe.

ANTIMATTER

Science fiction is positively bristling with antimatter. It dreams up vast quantities of the stuff.

Antimatter is made of stuff exactly like the stuff in ordinary material, but with the opposite charge. The idea was first mooted by physicist Paul Dirac in 1928. The existence of the positron, or anti-electron, was confirmed two years later.

Star Trek's Chief Engineer, Scotty, uses frozen anti-hydrogen as the primary fuel for the propulsion of the starship *Enterprise*, while physicists in Dan Brown's *Angels and Demons* (2000) manage to create enough antimatter to blow up the Vatican. And science fiction writers have imagined antimatter galaxies and even entire antimatter universes.

Antimatter cannot easily exist in our Universe. So far only trillionths of a gram have been isolated in labs, but the potential of antimatter is colossal. It would make a fantastic power source (if you could control the risks).

When antimatter collides with matter, the result is 100% mutual annihilation, with all of the mass being converted into pure energy. Einstein's famous equation $E=mc^2$ tells us that it only takes a very small amount of matter to create an enormous amount of energy. A mass of antimatter equivalent to a large car could produce all of the world's electricity for one year.

No wonder Scotty always looks worried.

➋ One giant leap for dogkind: Laika the space dog.
➌ The Earth is surrounded by an ever expanding shell of junk, every piece a tiny satellite.
➍ Scotty looking worried...

SUBTERRANEAN LIFE

Hell. The centre of the Earth. The locus of the Devil and his legions. The lowliest and most corrupt place in the entire Universe.

When considering our Universe, Aristotle had pictured an unchanging cosmos of nested crystalline spheres each made of ether, the fifth element. The lowly Earth was placed at its centre, the only region of space subject to change and decay.

The Church had gone a step further. Dante's great poem, the *Divine Comedy* (1606–21) describes a journey through the Christian Universe. The quest starts on the planet's surface, descends into the bowels through the circles of Hell, each populated by the sinful dead, and ends at the Earth's core. Dante's subterranean vision mirrored Aristotle's Universe above.

The notion stuck, for some at least, until the eighteenth century. One theologian suggested that the Earth's rotation was a result

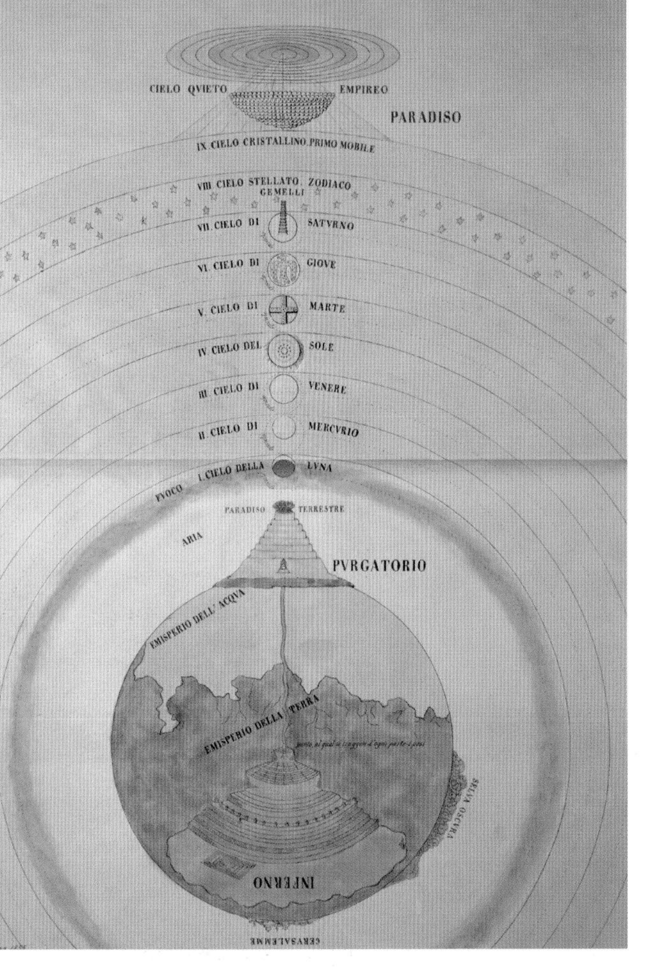

of the damned scrambling to escape Hell. The idea of nested spheres persisted in scientific circles. Newton's friend, astronomer Edmond Halley, proposed in a paper published by the Royal Society in 1692 that several rotating globes inside the Earth caused its magnetic field.

The first striking use of Halley's idea came in the form of Ludvig Holberg's *Nicolaii Klimii iter Subterraneum* (1745). Translated as *A Journey to the World Underground*, a young Norwegian stumbles down into the Earth to discover an inner planet populated by intelligent non-human lifeforms.

But the Industrial Revolution changed all that. The steam engine opened up the veins of the world. The fossil record spewed out signatures of previously unimaginable creatures and the dinosaur was discovered.

As geology advanced and the roll call of extinction grew, the terrible extent of history began to dawn. Jules Verne's *Journey to the Centre of the Earth* (1864) is a voyage through a subterranean world, and a conquest of space. Gone are Dante's mythical speculations of an Earthly core. In its place is a quest into the depths of evolutionary time. The explorers find the interior alive with prehistoric plant and animal life, and their aim is to possess nature for science.

When Darwinism provided writers with the concept of evolution, around seventy futuristic fantasies were spawned in England alone between 1870 and 1900. One of the first, Edward Bulwer-Lytton's *The Coming Race* (1871), was a book about subterranean supermen.

His fascinating, if bizarre, tale is set in a subterranean world of well-lit caverns. As with Ludvig Holberg's story, Lytton's book begins as the narrator falls into an underground hollow. Their heroes, it seems, are unduly careless.

Nonetheless, a mysterious human-like race is discovered, which derives immense power from *vril*, an all-permeating fluid that has enabled them to master nature. Indeed, *The Coming Race* proved to be truly inspiring for Scotsman John Lawson Johnston; he made a fortune from a strength-giving beef extract elixir known ever since as Bovril.

But the idea of subsurface life is with us still, on Earth and on Mars. The relatively new field of 'deep biology' has unearthed bacterial spores trapped in three-billion-year-old rock in

Dante's *Divine Comedy* is a journey into the bowels of Hell.
This dark spot on the surface of Mars is probably a skylight onto a subterranean cavern.
Architect Paoli Soleri.
Mega-City One from the Judge Dredd stories in *2000 AD*, an arcology with over 400 million residents.

a South African gold mine, and minute single-celled organisms – *foraminifera* – living at a depth of seven miles in the Marianas Trench, Earth's deepest point, in the Pacific Ocean.

And in 2007 NASA discovered seven candidate skylight entrances into subterranean caverns on Mars. All seven are located on the flanks of Arsia Mons (southernmost of the massive Tharsis-ridge shield volcanoes), a region with widespread collapse pits that may well indicate an abundance of subsurface void spaces.

Could there be some form of life in these Martian voids?

ARCOLOGIES

What do you get when you combine architecture with ecology? An arcology – an enormous habitat which people never leave. Effectively a self-contained environment, it is championed by the Italian architect Paoli Soleri. His 1973 book *The City in the Image of Man* argued that an arcology is more efficient and environmentally friendly and would enable our **BUILT ENVIRONMENT** to be ready for any future apocalypses.

Soleri acknowledges that science fiction came to this conclusion first. William Hope Hodgson offers a realistic vision in his 1912 novel *The Night Land*. The idea was pursued with Robert Silverberg's Urban Monads from *The World Inside* (1971), the city of Zion in *The Matrix* Trilogy and The Mega-Cities of *Judge Dredd*. These crime ridden megapolises are metropolitan slums that offer a cradle-to-grave existence for the citizen of the future. You can explore the concept in computer games like *Sim-City 2000* which allows you to follow Soleri's example by designing your own arcology.

The movement has already begun. You can already live in a small-scale Soleri-designed arcology – *Arcosanti* – in Phoenix. You will have to wait until 2040, however, when the new Chinese eco-city of *Dongtan* near Shanghai is finally finished to live in a large-scale arcology for real.

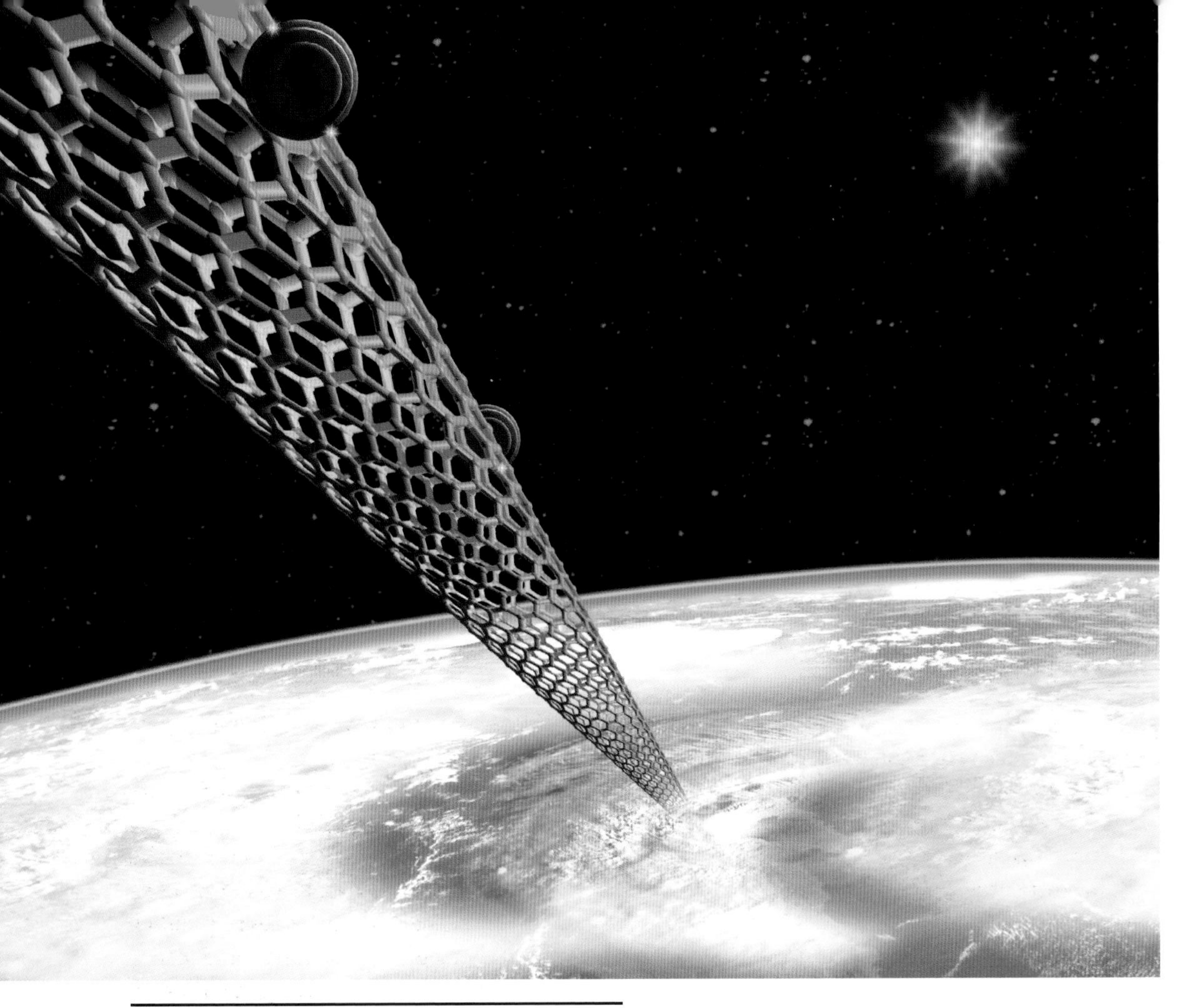

SPACE ELEVATORS

Imagine jumping into an elevator and pressing the button marked 'space'. That was the vision of Konstantin Tsiolkovsky who in his 1895 work *Daydreams of Heaven and Earth* speculated on such an event. It seemingly provided an ideal solution to **SPACE TRAVEL**. The idea was subsequently developed into a rigorous scientific proposal by fellow Russian Yuri Artsutanov in 1960 as a cheap and reliable method of getting large volumes of material into space. From that point on it has been considered by various groups and space agencies as a solution to both building **SPACE STATIONS** cheaply and encouraging **SPACE TOURISM**.

At its heart all the designs employ a 47,000 kilometre cable made from an as yet undeveloped composite material. This 'beanstalk' material would need to be thirty times stronger than steel and have a diameter of no more than ten centimetres so it could be tethered to an orbiting space station to provide continuous transport into space. Despite serious interest from NASA (the agency has spent millions of dollars running a competition on space elevator designs) the space elevator has never captured the public's imagination in the way that other science fiction innovated inventions have. Perhaps because, although popularised by Arthur C. Clarke in his 1971 book *The Fountains of Paradise*, it has made few appearances in film and TV series.

➊ 'Third floor: mesosphere; fourth floor: Karman Line ... Going up.'
➋ Simulating space life in SpaceStationSim.
➌ A 1957 simulator used to train astronauts for space travel.

SPACEFLIGHT SIMULATORS

Ever since humanity took to the skies we have been trying to find a way to practise flight effectively. First, entire aircraft were placed on a pivot to allow a pilot to practise on the ground. Then the famous 'blue boxes' were developed by aviation enthusiast Edwin A. Link to partially simulate a cockpit experience. But flight became more complicated. During the Space Race, travel in **ROCKETS** increasingly became too complicated an experience to replicate with mechanisms alone. The most effective way of learning to fly in space was by doing it.

Murray Leinster, in his 1953 short story *Space Tug,* had the answer. There his hero practises in a virtual simulator that replicates every sight, sound and feeling which **SPACE TRAVEL** could possibly throw up. Leinster had his heroes responding to staged emergencies, calamities and catastrophes in ways that the latest generation of spaceflight simulators are only now starting to be able to achieve. The movie *Apollo 13* (1995) features an early simulator being used to try to bring the stranded astronauts home. Things have come so far from Leinster's vision that you can now buy a video game – Space-StationSim – based on the actual program that NASA uses to train its astronauts for duty on the International **SPACE STATION.**

QUANTUM UNIVERSES

Is this the best of all possible worlds?

The question was first hit upon by German philosopher Gottfried Leibniz in an attempt to solve the problem of evil. Clearly not one to shirk the difficult questions, Leibniz' thinking went something like this: if God is good,

omnipotent and omniscient, how come there is so much suffering and injustice in the world?

Leibniz' solution in some ways pre-empted a science fictional obsession. He made God a kind of 'optimizer'. According to Leibniz, God simply chose from a host of all original possibilities and, since God is good, this world must be the best of all possible worlds.

Then came quantum mechanics.

The notion that our Universe is merely one of many is a facet of the 'many-worlds interpretation' of the mysteries of quantum theory. Propounded by physicists such as Hugh Everett and Bryce DeWitt, and popularised by writers like Paul Davies, the theory imagines an infinite number of parallel universes, making up a 'multiverse' that comprises all of physical reality.

Not only that, but such a multiverse contains all possible Earthly histories and all possible physical universes. Head hurting? Don't worry; it's natural. In fact, quantum supremo John Wheeler once said, 'If you are not completely confused by quantum mechanics, you do not understand it'.

In science fiction, such parallel universes may also be called 'other dimensions', 'alternate universes', 'quantum universes', 'parallel worlds', or even 'alternate realities'. Indeed, the idea that other worlds lie in parallel to ours is one of the oldest in speculative fiction.

At first, authors were slow to realise the potential extravagance in all this. The most distinguished exception, of course, is Lewis Carroll's 1865 story *Alice's Adventures in Wonderland*, in which the heroine pops into an alternate reality via a rabbit hole.

A notable early attempt to describe a parallel world with the same physics, but mapped onto fewer dimensions, is Edwin Abbott's *Flatland* (1884). Abbott's reality is one in which there are only two dimensions, rather than the usual three. Of course, the notion of **TIME AS THE FOURTH DIMENSION** would be popularised by H.G. Wells a decade later in *The Time Machine*.

Phillip Jose Farmer's short story *Sail On! Sail On!* (1952) describes an alternate 1492AD in which the Earth *is* flat. This other-world physics is such that Columbus sails over the edge of the world into Earth's orbit, never to return from his mission. Simply brilliant.

To be sure, Phillip Jose Farmer is something of a parallel world mastermind. Another of his tales seems like a whimsical perspective on Leibniz' original idea. *The Unreasoning Mask* (1981) features a multiverse, with each universe contained within a different cell of the body of God. The only means of travel between alternate universes is through the cell walls, which is wounding to the growing body of the still infant God.

And who can forget the movie ending of *Men in Black* (1997)? The final sequence reveals our Universe, held in a container resembling a marble. An **ALIEN** hand picks up the 'marble' and pops it into a bag full of universe-marbles!

Cosmologists David Deutsch and Max Tegmark are among the theoreticians who

Discrete 'bubble' universes – ours may not be the only one...
Jules Verne transports us from Earth to the Moon.
An infamous German V2 Rocket – the first object into space.

believe that quantum universes could actually exist. In a 2003 issue of *Scientific American*, Tegmark speculated that our Milky Way galaxy has a twin, in which there is a twin Earth, which in turn contains a twin of you.

SPACE TRAVEL

Imagine sailing your way into space. This was the vision of the satirist Lucian of Samosata. His 160AD *True History* (patently intended to be anything but) found the author and friends thrust into space by a massive waterspout whilst on a voyage of exploration. Not so much a piece of science fiction as a satire on great exaggerators like Homer, it took another fourteen hundred years before people started to explore space travel in a more sustained way.

The famed astronomer and 'father of planetary motion' Johannes Kepler wrote in his 1634 *Somnium* of a hero being drawn into space by spirits. It was left to a bishop, Francis Godwin, to come up with a physical approach to space travel in his story *The Man in the Moone* (1638), which features a mechanical means of reaching the Moon (although admittedly powered by geese!). It fell to Cyrano de Bergerac to first make use of **ROCKETS** and suggest a man could be propelled upwards in 1657. All of these stories, however, assumed that the Earth's atmosphere extended to the Moon and beyond. Numerous stories with unlikely devices for space travel followed until the invention of the balloon revealed to the scientifically inclined that high altitude travel is dangerous.

Conjecture paused for almost a century to be taken up by Jules Verne in his 1865 *From the Earth to the Moon*. Here the rudiments of space travel were in place. A sealed capsule was used to take the astronauts into orbit, the projectile being fired from a large gun. Variations on the sealed capsule theme are then pursued by other writers with more scientifically robust propulsion schemes being employed.

It was the father of Russian rocketry Konstantin Tsiolkovsky (who conceived the **SPACE STATION**, the **SPACE ELEVATOR** and **SATELLITES**) who provided a rigorously scientific approach to space travel. His 1903 paper *The Exploration of Cosmic Space by Means of Reaction Devices* set out this realistic vision. Tsiolkovsky was little known outside of Russia so it was an American, Robert Goddard, who proved more influential in the development of the rocket as the means of travel into space. Goddard's 1919 paper *A Method of Reaching Extreme Altitudes* was followed by a practical test in 1926 when he launched a rocket 41 feet into the air, before decimating his Aunt Effie's cabbage patch when it fell back to Earth.

This early work influenced two key European pioneers, Hermann Oberth and Wernher von Braun. It was from the latter's German rocket program that the first object made it into space. On 3 October 1942 a V-2 rocket was successfully launched beyond the Karman Line, the 100km limit that demarks the beginning of space. Space travel had taken a major leap forward.

From there the post-war Soviet regime took up the baton. The space race to launch a satellite resulted in the successful orbit of Sputnik One in 1957. Just four years later Soviet Cosmonaut Yuri Gagarin climbed aboard Vostok One and made a single orbit of the Earth. The dreams of mankind had only taken one and a half millennia to become a reality. Rockets were the main method of travel, although the re-usable space shuttle has now been deployed by the USA for just over quarter of a century.

Since these early expeditions, the idea of space travel has become even more popular. Yet by 2007 fewer than five hundred people had travelled beyond the Karman Line. All that is set to change. With the increasing interest in **SPACE TOURISM**, numbers could double within the next decade. Who knows, you could eventually find yourself reading this book whilst orbiting the Earth.

WEIGHTLESSNESS

The idea of weightlessness was first proposed by a science fiction Bishop.

In 1638, Francis Godwin, Bishop of Llandaff in Cardiff, published a story in which his hero hitches a ride to the Moon from a flock of strange birds called 'gansas'. At a certain point in the journey, the hero describes how 'neither I, nor

⬇ Professor Stephen Hawking enjoying weightlessness aboard a Zero Gravity Corp. Boeing 727 jet.

➡ The dawn of space tourism? An artist's impression of SpaceShipOne in re-entry. The craft was designed by US company Scaled Composites.

the Engine moved at all, but abode still as having no manner of weight.' Around the same time, German astronomer Johannes Kepler foretold the law of gravity. He believed the Sun exerted some mysterious power, or 'virtue', compelling the planets to hold to their orbits.

Taking Godwin's idea further, French science fiction writer Jules Verne in his novel *From the Earth to the Moon* (1867) correctly predicted that travellers in a spaceship on their way to the Moon would feel a weightless state: '... by the neutralizations of attractive forces, produced men in whom nothing had any weight, and who weighed nothing themselves.'

Weightlessness, however, is not only achieved at the point in the journey where the Earth's gravitational pull and the Moon's pull cancel each other out, as Verne suggests. Space shuttle astronauts in orbit mere miles from Earth also achieve weightlessness and, as Professor Stephen Hawking knows, passengers of the 'Vomit Comet' – a special aeroplane that is used to train astronauts – can also float in near zero gravity conditions as their plane rushes towards Earth in a ballistic parabolic dive.

SPACE TOURISM

Many people have dreamed of going into space but only a few hundred professional astronauts have so far made the trip. This may soon all change as private companies plan space tourism projects. 2001 saw the first official fee-paying space traveller, although space tourism had been much discussed in science fiction and science fact for some time before that.

Arthur C. Clarke's 1961 novel *A Fall of Moondust* describes a long-term commercial exploitation of space. Set on the Moon, this disaster story features a sinking sightseer transport in the midst of a lunar amusement park designed to entertain tourists. Carried to the Moon by shuttle, the tourists enjoy all that the lunar development has to offer. A similar vision was briefly outlined in Stanley Kubrick's *2001: A Space Odyssey* (1968), which he co-wrote with Clarke. In the film, a routine shuttle ride with tourists being served drinks by a stewardess is depicted as the movie's second sequence.

These stories preceded fee-paying tourism by some forty years, although some argue that in reality space tourism has been with us almost as long as the space program itself. By this they are referring to the differentiation between professional astronauts and payload specialists. These crew members do not undergo the same level of training and preparation for travel as the regular astronauts. This process began in 1963 with the first woman in space, Valentina Tereshkova, and has continued since that time. Often the representatives of commercial partners and companies, they are reluctant to be defined as space tourists because of its pejorative connotation and, not being self-funded, they probably do not fulfil the correct criteria anyway. Notable examples of payload specialists have been the journalist in space, Toyohiro Akiyama, who was sent by TBS Japan in 1990 and the first schoolteacher to set off to space, Christa McAuliffe, who perished in the 1986 Challenger disaster before reaching orbit.

In 2001 the debate about whether space tourism had yet begun was finally over when Dennis Tito, an American billionaire, travelled to the International **SPACE STATION**. Although Tito did carry out some experiments and was reluctant to be labelled a tourist, the self-funded nature of his journey leaves little doubt about his status in the eyes of the public. This opened the door and four further tourists (all multi-millionaires) have since travelled into space.

The possibility of mass transit remained elusive until the Ansari X-Prize was finally collected on 4 October 2004, the 47th anniversary of the launch of Sputnik, the world's first artificial **SATELLITE**. This challenged companies to design a re-usable method of taking a person into space, returning them to Earth and re-launching again all within two weeks. With the prize won and the technology within reach, numerous companies are developing facilities to offer regular flights into space before the first decade of the twenty-first century is out.

Of course, there is nowhere for them to visit. These flights would take the fee-paying passengers just above the 100 km barrier that – it is generally agreed – separates Earth from space, and then return them. No space hotel sits in orbit with which they can dock to enjoy some rest and relaxation. At least not for now. Once again commercial companies are anxious to exploit the possibilities of staying in space. Plans by Bigelow Aerospace for an inflatable space hotel mix with those of circumnavigating the Moon to offer the next generation of mud-dwellers on Earth the chance to soar to the stars.

Re-entry shields

Whenever you watch pictures of a spacecraft on re-entry its underside takes on an unearthly orange hue. This is the result of the heat caused by air compression. If the craft didn't carry a re-entry shield then the mission would be fatal. **SPACE TRAVEL** is inherently dangerous and barbequing yourself on the final leg of a mission is not a preferred option. Astronauts practise in **SPACEFLIGHT SIMULATORS** for re-entry in particular.

E.E. 'Doc' Smith, the acclaimed golden age science fiction author, recognised this hazard long before science caught up. His 1933 novel *Triplanetary* features the use of ablative shielding to combat the extreme heat caused by atmospheric friction. Mentioned as being sponge-like, his material, 'which air-friction would erode away, molecule by molecule, so rapidly that no perceptible fragment of it would reach ground', would save the occupants of the vessel.

This technology only started to develop following World War II as a consequence of both ballistic missile technology and the burgeoning space race. Early efforts were single use only, with the material being denuded during its passage through the atmosphere. The technology has now advanced so that the space shuttle features a reusable Thermal Protection System, although it was the failure of this that caused the 2003 Colombia disaster.

❶ The 1933 classic *Triplanetary* included a description of a re-entry shield.
❷ A typically understated branch of Carrefour.
❸ Theoretical physicist Freeman Dyson.

HYPERMARKETS

If we want to go shopping, rather than purchasing online, we are likely to visit a hypermarket. A combination of a supermarket that sells food and a department store selling general products, the hypermarket has become a one stop shop for everything the modern consumer could possibly want. The first hypermarket was opened in France by Carrefour in 1963 at Sainte-Geneviève-des-Bois. The key is to maximise the use of space, to 'pile it high and sell it cheap'. However, it was in 1888 that the idea of a hypermarket was first mooted in Edward Bellamy's science fiction novel *Looking Backward*.

Bellamy's vision was not an idealised image of the future. The picture he painted in 1888 was of a future dominated by vast state-controlled retail concerns forcing small businesses out of the market and restricting customer choice. He claimed that the former owners of the replaced shops and stores would then be forced to work as clerks in the 'great city bazaar' itself. These concerns monopolise the market until very few are left and consumers have run out of options. Sound familiar?

ARTIFICIAL ENVIRONMENTS

The most successful video game product of all time is not *Tomb Raider* or *Grand Theft Auto*; it is the science fiction first-person shooter called *Halo*. This game saw the player take on the role of a **CYBORG** soldier out to save the universe. Central to its conceit is the idea of the Halos themselves – huge ring-shaped **SPACE STATIONS** which orbit planets and have whole civilisations on their inner surface.

This idea can be traced back to the work of Freeman Dyson whose 1959 scientific paper *Search for Artificial Stellar Sources of Infra-Red Radiation* appeared in *Nature* and proposed orbiting structures, either spheres or swarms, designed to catch and collect all the energy radiating from a star. The original concept was for a swarm of **SATELLITES** to collect energy. Dyson reasoned that, inevitably, any growing civilisation would require so much energy that its only option would be to surround and harness the power of a star directly. His paper was sketchy on the mechanics and focused instead on the energy conversion process itself. This process, Dyson suggested, would alter the wavelength of the energy emitted by the star and this change would allow astronomers to discover any spheres or swarms that already existed. Here, Dyson reasoned, may lie proof of alien life. Indeed, the Search for Extraterrestrial Intelligence (SETI) program has a search for Dyson spheres as one of its tasks.

The Halo is a slice of Dyson sphere in virtual form. Yet, just as the Halo is science fiction based on scientific research, so that scientific research was itself based on science fiction. Only this time it was not a game but a book, *Star Maker* (1937) by Olaf Stapledon. In this hugely influential book, the British narrator explores the universe and in the process comes across objects very similar to Dyson Spheres. Dyson explained that he read Stapledon many times and took the concept of his thought experiment from its pages. Science fiction drove the science forward, which in turn inspired science fiction to new heights of creative endeavour.

Following Dyson's work, his concept has made the transition into popular science fiction. In books like Larry Niven's *Ringworld* (1970) and Arthur C. Clarke's *Rendevous with Rama* (1972) the plots are played out on the inner surface of slices of sphere. The *Star Trek: The Next Generation* episode 'Relics' features a Dyson sphere. After being trapped inside, the crew try to escape the inner star's solar flares which threaten the ship. It's up to the two engineers Lt. Geordi Laforge and Chief Engineer Montgomery Scott (from *The Original Series*) to save the day. This is done by wedging open the Dyson sphere's doors with Scotty's ship and having the *Enterprise* blast through with photon torpedoes at the last moment. Scotty had effectively engaged in **TIME TRAVEL** by being electronically stored in the **TELEPORTATION** device, the transporter.

In the episode it is a solid shell from which the heroes escape. However, following on from Dyson's research, the shell is one of only four variants of Dyson sphere. The original paper proposed a 'ring' effect with a loop of satellites around the star. This was then developed with multiple rings of satellites coming together to form a Dyson Swarm. A further model is the Dyson Bubble which similarly features evenly spaced collectors around the star although this time not satellites but 'statites'. These statites hang in place and do not orbit the star at all. Finally the sphere itself, often referred to as a Dyson shell, would be by far the most complicated to construct, and yet is also the most commonly depicted in fiction.

One of the ring-shaped artifical environments from *Halo: Combat Evolved*.
Bruce Jensen's artwork for Neal Stephenson's 1992 bestseller *Snowcrash*, which features Earth, the virtual globe program.
Moonraker, James Bond's space jaunt.

VIRTUAL GLOBES

One of the internet's most popular sites is *Google Earth*, launched in 2005. This allows you to zoom in from a virtual globe right down to your back garden, observing

the changing features of the Earth as you fly. This latest instance of a virtual globe (Norkart's 2001 *Virtual Globe* and NASA's 2004 *World Wind* are other examples) has a science fiction origin.

As is the case for **AVATARS,** Neal Stephenson's cyberpunk book *Snowcrash* is the source. It featured a program developed by the Central Intelligence Corporation simply called 'Earth'. This allowed the user to navigate around the planet. Swooping around you could home in on any facet that you wanted to examine in more detail. Cinematically, Will Smith starred in the 1998 science fiction thriller *Enemy of the State* which used **BIG BROTHER** sequences similar to a real-time virtual globe to allow the National Security Agency (NSA) to trace the hero. The television series *Star Trek: The Next Generation* and *Stargate: Atlantis* have both gone further with the creation of virtual universes allowing virtual characters to live out virtual lives whilst being monitored from the real world.

SPACEWALK

Whenever something is assembled or repaired in space someone has to leave the safety of their vehicle and venture outside. Formally known as Extra Vehicular Activity (EVA), it is more commonly called a spacewalk.

It was the science fiction illustrator Frank R. Paul's prescient cover image for the fall of 1929 edition of *Science Wonder Quarterly* magazine that set the stage. In it, pressure-suited astronauts, held by lifelines, float around a **ROCKET** experiencing **WEIGHTLESSNESS.** One of the astronauts is depicted using a guidance gun later developed by NASA in a compressed air version and used on the Gemini 4 mission of 1965.

The first spacewalk was carried out by Soviet cosmonaut Aleksei Leonov in 1965 and this practice continues today. Spacewalks have appeared in numerous films including James Bond's jaunt in *Moonraker* (1979) (which also

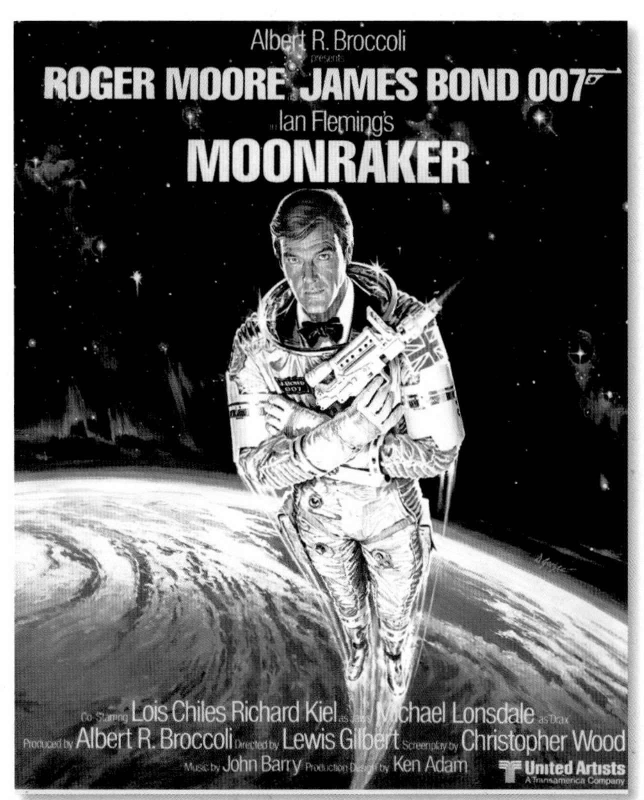

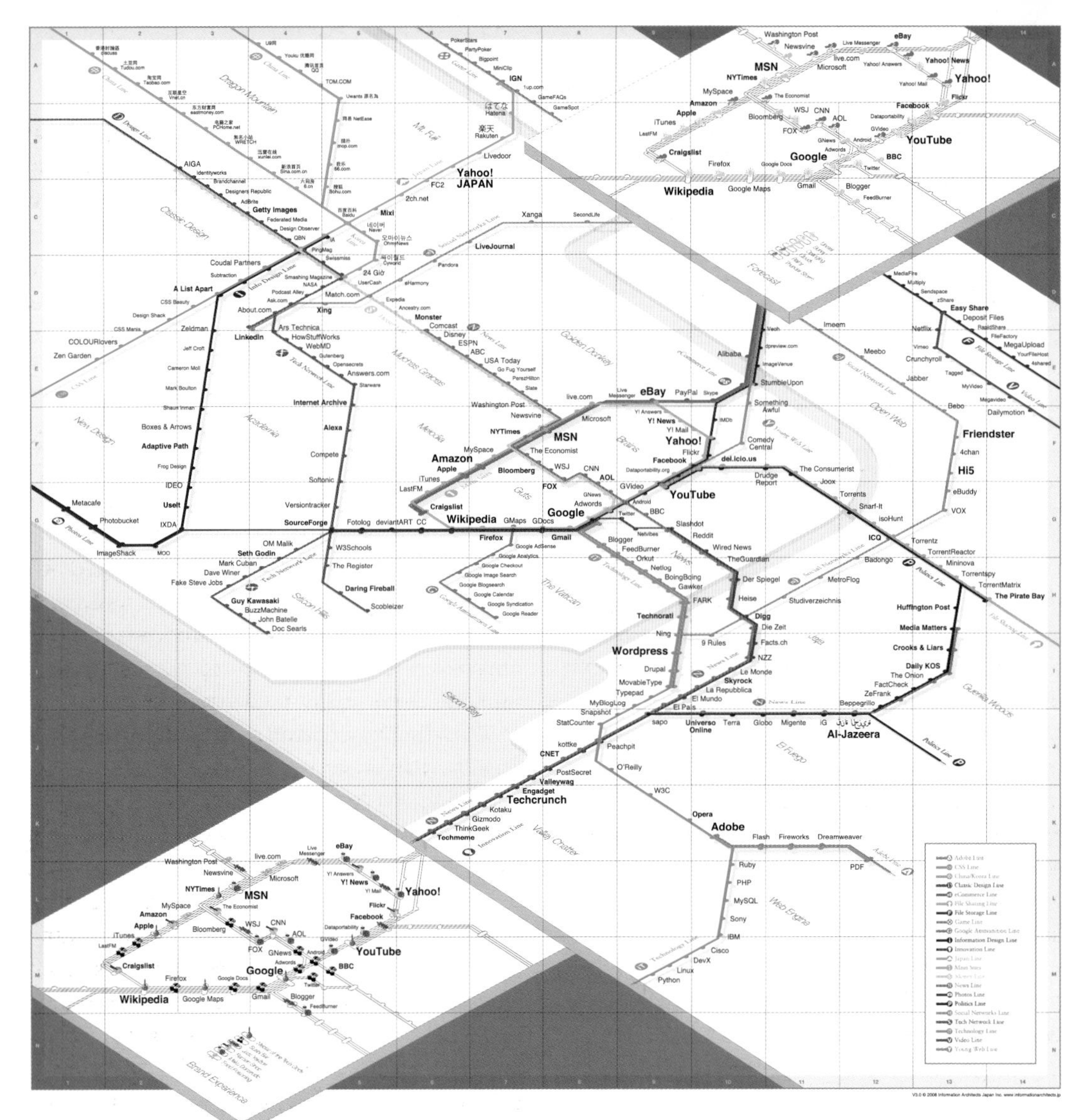

showed Space Shuttles three years before they were launched) and the sequel to *2001: A Space Odyssey*. In *2010: The Year We Make Contact* (1984) John Lithgow makes a wonderful spacewalk in Jupiter space.

Although a suit was punctured during a NASA spacewalk (in 1991) it was not fatal. This has still led to continuing concerns over the safety of spacewalks and other options (including **ROBOTS**) are being explored.

CYBERSPACE

On 27 April 2007 Russia declared war on Estonia. At least that's what the Estonian Government claimed. This was not conducted on land, at sea or in the air. It was instead conducted in cyberspace. The 2007 Estonian Cyberwar saw electronic warfare conducted across cyberspace against Estonian banks, media outlets and the government. It was claimed by Estonia that this was a state-sponsored attack from Russia that was meant to cripple the Estonian economy. Cyberspace just got more dangerous.

But what is this elusive cyberspace? It has grown to dominate the modern world and evidently to provide not just a place to work, rest and play but also to wage war. It was the influential science fiction author William Gibson who first coined the term in his 1982 short story *Burning Chrome*, in which it is described as a portmanteau of cybernetics and space.

A map of cyberspace: the Web Trend Map puts 300 websites on the Tokyo subway map, linked by theme and positioned by popularity.
Author William Gibson, who coined the term cyberspace.

This tale, set in a dystopian future dominated by electronics, has been labelled as part of the cyberpunk sub-genre of science fiction which developed in the 1980s. Gibson not only introduced the term but also gave us one of the most enduring descriptions of cyberspace. In his hugely influential 1984 novel *Neuromancer*, Gibson describes this elusive concept as a 'consensual hallucination'. In the book the hero, Henry Dorsett Case, is a console cowboy or hacker who is retained by a mysterious benefactor to perpetrate a seemingly impossible crime. However, all is not what it seems. Case penetrates this construct which exists seemingly nowhere, yet everywhere. Another science fiction author, Bruce Sterling, describes cyberspace as the place where your conversation on the telephone *appears* to take place, a conceptual void established by the participants of the call.

The first mention of this sort of place was with Plato's 360BC *Republic*. In its seventh book Plato creates the fiction of 'The Cave', an allegorical place where prisoners debate the nature of reality. It seems completely real to them, yet it does not exist. By extension, when we enter cyberspace by logging on to the **INTERNET** we are entering a realm which both does and does not exist. Although we are physically connected to a diffuse network of computers, we seem to have entered an alternate reality. Cyberspace has become a synonym for our explorations across the World Wide Web. The websites and bulletin boards that we encounter can be said to exist in Cyberspace. Thus the untouched or unviewed webpages exist in some kind of electronic limbo waiting to be inspected. As we typed our way across this brave new world we wanted to immerse ourselves still further and so we used **AVATARS** to insert virtual representations of ourselves into this non corporeal realm.

Initially these were used just to communicate with other users. To provide handles by which we became known. More recently, however, these avatars have been provided with their own playgrounds, incarnations of cyberspace where they can live and play. Cyberculture is starting to provide a genuine alternative to our own existence. People started to establish virtual communities. Initially they focused on online role-playing games. Soon, however, the plot elements of the game came to be replaced with the ultimate narrative – life itself. Sites like *Second Life, Red Light Center* and *Millsberry* all provide users with a wide variety of social interaction via their avatars. Already commercial interests have meant that several individuals have become real life millionaires by virtue of their virtual existence. Perhaps its time for you to take a ride into cyberspace.

ANTI-GRAVITY

One of the great science fiction dreams. The idea of a force that opposes gravity emerged in the late nineteenth century. Typically, writers imagined devices allowing people, or objects, to hover or to be boosted about. An anti-gravity principle, known as

'apergy', was used to send spaceships to Mars in Percy Greg's *Across the Zodiac* (1880). Less romantically, in C. C. Dail's *Willmoth the Wanderer* (1890), an anti-gravity ointment is smeared on the hero's space vehicle.

Unsurprisingly, we owe the most famous antigravity device to H.G. Wells. His *First Men in the Moon* (1901) ingeniously describes how anti-gravity shutters made of 'Cavorite', a metal that shields against gravity, is used to send rockets to the Moon. And Buck Rogers even had an anti-gravity belt.

Since, according to Einstein, gravity is just curved space, all that's needed to allow antigravity is to simply bend space the other way. Tricky. For an anti-gravity device to work, 'negative mass' would be needed, an idea only conceivable in a universe of 'negative space', which could not exist in our own Universe.

Nonetheless, US physicist Paul Wesson and his colleagues are re-considering Einstein. And in 2011, they hope to launch an experimental satellite to see if 'negative mass' could exist.

SPACE STATIONS

The first space station ever mentioned was the result of an accident. The 'brick moon', conceived by Edward Everett Hale in his novel of the same name, was supposed to be a **SATELLITE** which served as a navigational aid for travellers. Instead in the 1869 story it is accidentally launched with passengers aboard. It was followed by other science fiction stories including those by Konstantin Tsiolkovsky (who also conceived the **SPACE ELEVATOR**). Tsiolkovsky set out to explore the scientific feasibility of space stations in his professional work as a direct consequence of the inspiration provided by fiction.

It was the German rocket scientist, Hermann

Artwork from *2001: A Space Odyssey* – a typical spoke design space station.
A city on the Moon: an artist's impression of the first step towards space colonisation.

Oberth, who first used the term 'space station' in his 1923 *The Rocket into Planetary Space*. He employed it to describe the wheel-shaped, spoke-filled habitat that he believed would serve to transit passengers to Mars. It was Oberth's pupil, Werner von Braun, who, after an early career in Germany, moved to NASA and helped popularise the concept. As a consequence the spoke design made an appearance in the 1968 film *2001: A Space Odyssey* amongst others. It is a design which so far has not been made a reality. When the first space station Salyut 1 was launched in 1971 it and subsequent stations (including Mir and the International Space Station) followed a modular pattern.

SPACE COLONISATION

What is the future of the human race? Will we blow ourselves up? Will over-population come to threaten our very existence? Will there be complete environmental breakdown from climate change? As soon as any of these topics are mentioned in the pub, some smart alec always chimes in with, 'Well, we can always colonise space …'

The first time colonisation in space is mentioned it is the Earth which is a colony. Bishop Francis Godwin's 1638 classic science fiction story *Man in the Moone* has the King of the Moon shipping undesirables off to Earth. Our planet is seen as a giant penal colony filled with the outcasts of the Lunar utopia. *Brick Moon* from Edward Everett Hale features an

accidental **SPACE STATION**, although the intent is not to colonise space. Tsiolkovsky, the Russian science fiction author and scientist, explored colonising space (initially using **SPACE ELEVATORS** to get into orbit), conjecturing that mankind should build greenhouses to support such colonies. But where would we place them?

Planets seem initially to be a sound bet. Before finding 'other Earths' building in our own solar system seems to be the natural way forward. We would need solid ground to build upon and some gravity as well. One of the problems is the distance involved. Mars appears to be the current favourite although the travel time and expense involved make it a difficult prospect. Of even more concern are the effects of microgravity on the human body. Any colony on the Moon would suffer from a similar problem of reduced gravity, although the relatively close distance to the Earth makes travel a lot easier.

Scientists speculate that an orbital colony is probably the first step. With the International Space Station already in place and **SPACE TOURISM** driving forward plans for hotels in orbit, this seems a logical move. An orbital station can be made to spin, giving the illusion of gravity. It can be constructed on the Earth in modular sections and placed in orbit without having to manufacture it in off-planet hostile environments. It would receive the same sunlight as the Earth, allowing biodomes to grow food. These space cities could house hundreds or even thousands of people in the future. In the beginning, however, a self-sustaining colony would need just one hundred and fifty people to ensure an adequate genetic mix. Perhaps you could be one of those future colonists?

Science has already tried to emulate these types of colonies on Earth. The closed ecosystem which would have to be sustained for a self-supporting colony was trialled by Soviet scientists in the BIOS3 project at the Institute of Biophysics in Krasnoyarsk, Siberia, between 1972 and 1984. This restricted facility only has

Exterior view of the Biosphere 2 project in Arizona.
A UFO from *Independence Day* soon turns out to be a spaceship with unfriendly occupants.

space for three occupants for about six months and uses xenon lamps to simulate the Sun.

A more ambitious project was the Biosphere 2 project, a 3.14-acre structure originally built in Oracle, Arizona, between 1987 and 1991. In it a crew of eight spent two years sealed away from the world. They lived in a giant biodome which featured an 850 square metre ocean with a coral reef, 450 square metres of mangrove wetlands, a 1,900 square metre savannah grassland, a 1,400 square metre fog desert, a 2,500 square metre agricultural system and a human habitat with living quarters and office. Initially self sustaining, problems in producing enough food and oxygen meant that the environment had a similar feel to a human habitat at 4,000 feet above sea level. The Biosphere 2 project proved how difficult a space colony would be to sustain. Aside from scientific and technical problems the crew themselves fell out and the whole project was eventually sold in 2007.

UFOS (AS FLYING SAUCERS)

It's true that it was US businessman Kenneth Arnold who coined the phrase 'flying saucer' on seeing nine discs flying in formation in 1947. The infamous Roswell Incident of the same year centred on claims that a flying saucer had crash-landed in the New Mexico desert. But it was science fiction that gave us the spaceship. One year earlier, science fiction writer Theodore Sturgeon had penned a short story, *Mewhu's Jet* (1946), with striking similarities to the Roswell Incident. And time after time in the decades before Roswell, science fiction had presented the spaceship as cutting edge technology.

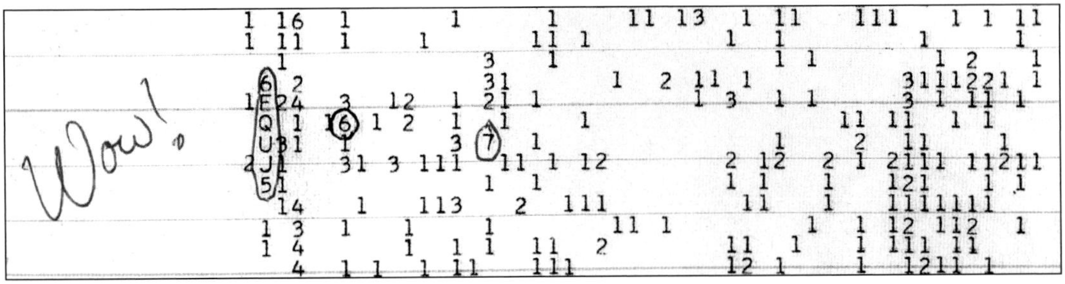

So it's no wonder that at the height of the Saucer Craze of the 1950s and early 60s, people came to the crazy conclusion of the 'ET hypothesis' – flying saucers were piloted by **ALIENS** from another planet.

Unexplained 'objects' in the sky are nothing new. Many phenomena have astronomical or meteorological explanations. 'Airships' were seen by passengers on ships crossing the Atlantic during 1896–7. Before that, witches on broomsticks were very popular. Sightings are interpreted using the believable technology of the day. And who knows, in these days of the Nimbus 2001 and the Firebolt, broomstick sightings may well return.

10 great sci-fi movie taglines

- 'In space no-one can hear you scream.' *Alien* (1979)
- 'Man has made his match … Now it's his problem.' *Blade Runner* (1982)
- 'Nice Planet. We'll take it.' *Mars Attacks* (1996)
- 'A long time ago in a galaxy far, far away….' *Star Wars* (1977)
- 'The last man on earth is not alone.' *Omega Man* (1971)
- 'The bitch is back.' *Alien*³ (1992)
- 'There is no gene for the human spirit.' *Gattaca* (1997)
- 'God made him simple. Science made him a god.' *Lawnmower Man* (1992)
- 'They stole his mind, now he wants it back.' *Total Recall* (1990)
- 'The future is history.' *Twelve Monkeys* (1995)

THE SEARCH FOR ALIEN LIFE

So deep is our conviction that there must be life out there beyond the dark, we half-expect aliens to land on Earth at any moment. Then, mulling over the immensity of time, we wonder whether contact came long ago. Perhaps a bright projectile plunged into the swamp muck of a steaming coal forest. Maybe a probe was clambered over by hissing dinosaurs, its delicate instruments running down with little to report …

If Darwin was right and, furthermore, if the principle of evolution reigns supreme in all worlds, how does man measure up?

The question clicked with Johannes Kepler almost as soon as Galileo discovered our Moon might be liveable. And H.G. Wells quoted Kepler at the very start of his Martian invasion: 'But who shall dwell in these worlds if they be inhabited? Are we, or they, Lords of the World?'

Wells created the myth of a technologically superior extraterrestrial intelligence (ETI). *War of the Worlds* features the 'men' of the future in alien form. They are what we may one day become. They are the tyranny of intellect alone and Imperial Britain, for once, is on the receiving end of interplanetary Darwinism. Wells destroys the idea that man is the pinnacle

↖ The Wow! signal, discovered by the Big Ear radio telescope in 1977, was possibly of extraterrestrial origin.
↘ Alvim Correa's celebrated take on Wells' tripods, 1906.

of evolution, though, curiously, the Martians were also complacent; earthly microbes eventually trounced them.

Are we alone in the Universe? The subject of alien contact and man's place on the cosmic evolutionary ladder became a twentieth-century obsession.

It wasn't until the 1960s that astronomers realised the potential shock of discovering extraterrestrial intelligence. However, writers like Olaf Stapledon and Arthur C. Clarke were twenty years ahead of the game. Through fiction they were preparing the public for a close encounter of the third kind: physical contact. It would be the final great demotion for human arrogance and the assumption that man is the measure of all things. Books like Stapledon's *Star Maker* (1937) and Clarke's *Childhood's End* (1953) stressed the insignificance of humanity in the face of alien biologies and ETIs.

As a result, scientists started searching for the alien. The emotional question of our place in the Universe was woven into all scientific discussions on ET. The fiction of alien contact had a profound effect on working scientists, such as exobiologist J.B.S. Haldane, physicist Fred Hoyle, and the founders of the Search for Extraterrestrial Intelligence (SETI) in the early 1960s, Carl Sagan and Frank Drake. Drake became the first radio astronomer to contemplate the transmission of an alien contact signal, greatly influencing the design of listening programs using the largest radio telescopes on Earth.

The highway of the scientific imagination has two-way traffic. Fact influences fiction, and fiction fact. For over 100 years, both had been firmly pro-SETI and pro-life in the extraterrestrial life debate. In the words of Arthur C. Clarke, 'the idea that we are the only intelligent creatures in a cosmos of a hundred billion galaxies is so preposterous that there are very few astronomers today who would take it seriously. It is safest to assume, therefore, that They are out there and to consider the manner in which this fact may impinge upon human society'. But in the last thirty years or so, things have changed. Pioneers of biology, such as Theo Dobzhansky and Ernst Mayr, have emphasised the incredible improbability of intelligent life having ever evolved, even on Earth.

Nonetheless, fiction has swayed an entire generation of future SETI-enthusiasts. It is astonishing that millions of dollars have been spent on sober scientific projects in the search for extraterrestrial intelligence. There can be no greater testament to the power of the scientific imagination.

ALIENS

The question of alien life has produced some of the best movie taglines of all time. And science fiction writers and directors have thought long and hard about the portrayal of creatures from other worlds.

The predatory and possessive mother in Ridley Scott's *Alien (1979)*, the swirling sentient sea in Stanislaw Lem's *Solaris (1961)*, and the hyperactive glove-puppet with poor sentence construction from George Lucas' *Star Wars* movie series. Alien as highly evolved killer, alien as ocean-planet, and alien as wise, benevolent, if slightly ridiculous, mentor. Science fiction, inspired by the findings of science, has been conjuring up aliens for many a moon.

Enthused by Galileo's discoveries with the telescope, German astronomer Johannes Kepler was one of the first writers to imagine alien life. The extraterrestrials that stalk Kepler's proto-science fictional book *Somnium* (1634) are not humans. They are creatures fit to survive their alien haunt. Two centuries before Darwin, Kepler had grasped the bond between life forms and habitat.

But mostly, before science fiction really began in the nineteenth century, extraterrestrials were not genuine alien beings. They were merely men and animals living on **OTHER EARTHS**. It was Charles Darwin who changed all that. For Darwin invented the alien.

Darwin's theory of evolution gave science fiction grounds for imagining what life might develop in space. From now on, the notion of life beyond our home planet was linked with the physical and mental characteristics of the true extraterrestrial. And the idea of the alien became deeply embedded in the public imagination.

The archetypal alien, with its strange physiology and intellect, owes much to H.G. Wells' 1898 Martian invasion novel *War of the Worlds*. Wells' Martians are agents of the void; they are the brutal natural force of evolution, and history's first menace from space.

Wells' genocidal invaders, would-be colonists of planet Earth, were so influential that the alien as monster became a cliché in the twentieth century. But the idea thrills us still. The alien as monster stalks the *Nostromo* in Ridley Scott's electrifying movie, and lies at the heart of each Dalek in *Doctor Who*.

With advances in biology, writers became more imaginative about alien life forms. Darwinism travelled into space with French astronomer Camille Flammarion in 1872. His three *Stories of Infinity* were ingenious tales of an intangible alien life-force.

If natural selection was universal, there was no reason on Earth why the random process of evolution should produce humans on other planets. Distinguished British astronomer Fred Hoyle used his science to inspire his stories. But his fiction was not forced by his physics. Hoyle's first novel, *The Black Cloud* (1957), is about a living cloud of interstellar matter.

Strange alien life forms reached their peak with Stanislaw Lem's famous *Solaris* (1961, with film versions in 1972 and 2002). Now an entire planet, Solaris, enclosed by an ocean, is a single organism with a vast yet strange intelligence that humans strive to understand.

Science fiction has also put forward the notion of wise, benevolent aliens who will save us from ourselves. In films such as *Close Encounters of the Third Kind* (1977), we are presented with civilised and benevolent aliens of superior intelligence, whilst aliens such as Yoda from *Star Wars* possess an almost saintly wisdom.

Even given the tremendous advances in the understanding of space during the twentieth century, science still has little to say about the psychology and physiology of the alien. But science fiction has been conducting thought experiments on the matter for centuries.

↑ The alien as hyperactive glove-puppet: Jedi Master Yoda.

→ A xenomorph created by H.R. Giger for *Alien* (1979).

What if we could tamper with time? What if we could jump timelines, guide evolution, re-write history, or even cheat death?

Earlier, we saw that science fiction is all about the relationship between the human and the non-human. And that the non-human is some form of the natural world, as revealed by science.

Ever since the Scientific Revolution, science has encroached upon all aspects of life. Science seeks not just to explore, but to exploit nature. To master it.

Around the seventeenth century the possibility that time was limitless and inhumanly vast in scale was put forward; even the stars, it was suggested, may grow old and die.

By the time of the Industrial Revolution, the death toll of extinction was becoming clear for the first time as the fossil record churned out evidence of creatures no longer to be found on Earth. Darwinian evolution forced us all to confront the terrible extent of history.

Suddenly, there was no greater challenge for science. What if we could master time, the brutal agent that devours beauty and life?

And so began a science fictional obsession. True, there had been folkloric flirtations with time. Time-slip romances where dreamy magic is mixed with myth, and time is lost as a convenient plot device.

But the idea of mechanised time travel did not appear until industrialisation. Its invention was tied up with the concept of time itself. The ancient Greeks had two words for time, *chronos* and *kairos*. *Kairos* suggested a moment of time in which something special happens. *Chronos* was more concerned with measured, sequential time. Industrial society brought a mechanistic approach to nature. *Chronos* came to the fore and the idea of time

H.G. Wells gave science fiction one of its most enduring devices – the time machine. Wells' book is an ingenious voyage of discovery. The Time Traveller sets out to marshall and master time. But he discovers the inevitable truth: time is lord of all. The full significance of the story's title becomes clear. Humans are trapped by the mechanism of time, and bound by a history that leads to inevitable extinction.

But science fiction writers continued in their quest to master time, to nail down the future for us.

Wells himself wondered what the future held for man in *The War of the Worlds*. The invading Martians are not only a brutal force of evolution, they are also the 'men' of the future. They are alien, yet they are human. They are what we may one day become, with their over developed brains and emaciated bodies. They are the tyranny of intellect alone.

Evolution, of course, is a process that reveals itself in time. In *2001: A Space Odyssey* Stanley Kubrick suggested that the dumb blind evolution of man would have to be interrupted by the guiding hand of an alien race to rescue us from the long, pathetic road to racial extinction.

But time travel proved to have further potential for storytellers: it enabled them to instill a different sense of history through the creation of plausible alternatives. A common theme was a Nazi victory in World War II, as in Robert Harris' *Fatherland* (1992), for example. The book is a counterfactual history of a single past created to provide long-term social and political speculations on history.

Philip K. Dick's classic *The Man in the High Castle* (1962) is also a Nazi counterfactual. But in Dick's book, rather than portraying just one timeline, some of the book's characters ar

is authentic? Which history is true? *The Man in the High Castle* helped develop alternative history fiction as a serious genre.

In film and fiction, it soon became common for characters to jump freely between alternate timelines, each timeline associated with its own plausible future. So it is with both the *Back to the Future* and *Terminator* series, the latter of which features apparently inescapable cyborg assassins sent back from the future by a race of artificially intelligent machines bent on the extermination of humans.

Inevitably, the main focus of such time traveling is melodrama. But these tales in time also enable us to ask questions: How open is the future? Do we really have free will? How can we ever know anything about time other than the fables we create?

TIME TRAVEL

You know how it is. A car comes out of nowhere, and slam! Before you know it, it's 1973. Are you mad, in a coma, or back in time?

Time travel in science fiction, such as that from the BBC's *Life on Mars*, is often genius. And so it may become in science fact.

It all started in 1895 with *The Time Machine*. H.G. Wells realised that if there's one thing in this big new Universe you can't tamper with it's time. So, he tampered with it. Figuring that space made up three dimensions, Wells volunteered time as the fourth. And if you can move freely in 3D, why not also in time?

Time travel happens every day, of course. Time's arrow moves us forward as we experience a sequence of events. Stargazing is a form of time travel. Our Universe is so unimaginably vast that light from its outer limits, picked up by our most powerful telescopes today, set off on its journey more than 13 billion years ago. The nearest star system, Alpha Centauri, is about 25 trillion miles away. Light, the fastest thing known to science, takes more than four years even from those nearby suns to hit the naked eye. So, we see the sky as it was in the past.

The truth is, if you think about this stuff long enough, your head can start to swim. Since nothing happens instantaneously, and light takes time to reach the human eye, even a ship's sail, for instance, emerging on the distant horizon is seen as it was an infinitesimal fraction of a moment ago. This phenomenon would be far more obvious if the speed of light was much slower than its 186,000 miles per second (300,000 km per second).

The idea of time travel in fiction is usually one in which the alleged illusion of time's arrow becomes (a) a confusion of time-hopping and causal relationships (*Groundhog Day*), or (b) a multiplicity of possible timelines (*Back to the Future*). DCI Sam Tyler, from *Life on Mars*, was not the first to be unstuck in time. Billy Pilgrim, the protagonist in Kurt Vonnegut's wonderfully creative anti-war classic *Slaughterhouse Five*, is similarly unstuck, and *Doctor Who* is founded on the idea that time is a Wellsian dimension that can be traversed at leisure!

Wells' treatment of **TIME AS THE FOURTH DIMENSION** came a full decade before Albert Einstein rewrote physics with his Relativity Theory. And Einstein's treatment of space and time as part of a four-dimensional 'spacetime' has exploded into today's conception of a multi-dimensional Universe. String theory predicts that spacetime has up to 10 or 11 dimensions, but that these are curled up in subatomic dimensions.

Another piece of time travelling genius from science fiction is Carl Sagan's *Contact* (1985). The famous US astronomer sought advice from **BLACK HOLE** expert Kip Thorne before making the film version of *Contact* (1997). Sagan wanted to know whether black holes could be used to travel in time. Thorne's advice: black holes are dangerous; use a **WORMHOLE** idea instead. Indeed, Sagan's enquiries about time travel galvanized a whole new field of physics, with dozens of scientific papers by some of the world's best physicists, including Kip Thorne himself and cosmologist Paul Davies who wrote *How to Build a Time Machine* (2001).

However, there are still technical issues before a wormhole time machine becomes a reality. We've not yet found a wormhole, let alone learned how to control one. But it's a stunning fact that, early in the twenty-first century, we have now reached a stage in our understanding of nature where time travel is even a bare possibility.

PRECRIME

The classic 1956 science fiction story *Minority Report* by Philip K. Dick features a precrime division of the Washington DC police force which uses a combination of precognitive psychics and computers to predict when and where crimes will take place. In the 2002 film version of the same name, Tom Cruise plays a hero cop who dashes from place to place desperately stopping pre-criminals before he is framed himself, effectively turning society into a **BIG BROTHER** state in which you are condemned for what you are going to do. In recent years, this bleak vision of the future has taken a significant step closer to coming true.

The New York Police Technical Department is pioneering work which uses computers to predict when and where crimes will occur. Its first success took place when it correctly analysed and predicted a street robbery would occur on 27 July 2005 between eight p.m. and midnight on South Broadway in Yonkers NY. As forecasted, a woman was robbed of her

◀ A young H.G. Wells.
▲ *Life on Mars*, a new take on time travel.

MOBILE PHONE within 44 minutes of the period starting. Officers who were cruising the area specifically because of the prediction were able to arrest the armed suspects immediately. The use of computers to help predict crime patterns is set to rise in the future when you will have to be careful about not only what you do but what you plan to do.

❶ Time Machinery – Guy Pearce is the Time Traveller in the 2002 film *The Time Machine*.
❷ 'The system is perfect. Until it comes after you.' Max von Sydow as Director Lamar Burgess in *Minority Report* (2002).
❸ Professor Ronald Mallett has used Einstein's equations to design a time machine.
❹ Fritz Lang's *Frau im Mond* (1929), the first time the countdown was used in a rocket launch.

TIME MACHINES

Your machine is built. Where to now? The last days of the Roman Empire? The Dark Ages? Or that terminal beach at the Heat Death of the Universe? The Dark Ages. Good choice.

Temporal camshaft. Check. Fourth-dimensional perambulator. Check. Ignition. Check. Ignition?! Hang on a minute; exactly how are these time machines meant to work?

One of the very first time gadgets was invented by Charles Lutwidge Dodgson, better known as Lewis Carroll. His idea: the Outlandish Watch. Appearing in Carroll's *Sylvie and Bruno* (1889), the watch had two modes. If the reverse peg is pushed, then 'events of the next hour happen in reverse order'. The other mode involved the watch's hands. They could be moved backwards, as much as a month, enabling the wearer of the watch to travel into the past.

You're not, of course, going to find one of these little beauties at the local jewellers. And the device's owner, a Professor for heaven's sake, is frustratingly elusive when asked to describe the theoretical basis for the thingamajig: 'I could explain it, but you would not understand it'.

Wells fared no better. He wisely left his time machine extremely fuzzy. It was left to others to go figure. French writer Alfred Jarry stuck his neck out in a review of Wells' *Time Machine* called *A Commentary to Serve for the Practical Construction of the Machine to Explore Time*. Deft little title, isn't it?

Wells had talked vaguely of levers. But Jarry unduly designed his machine with three rotating gyrostats. He even included an impressive nonsense diagram. Jarry figured that a time machine would have to anchor itself absolutely in space in order to move in time. In this way, he says, 'all future and past instants ... would be explored successively'.

Kurt Vonnegut's race of fictional **ALIENS**, the Tralfamadorians, had gone even further. With no contraption of any kind, they were naturally able to see along the timeline of the Universe. They not only had the ability to experience all four dimensions, they also had total recall of both past and future. The human view of time is a mere snapshot, whereas the Tralfamadorians' is that of a movie, in which all scenes are played out at once.

Einstein had died a Wellsian. He wrote in 1955 that enlightened physicists understood the division of past, present and future to be a mere illusion.

So, how feasible is Wells' fictional trip into the far future? Professor of Theoretical Physics Paul Davies addressed such questions in his 2001 book *How To Build a Time Machine*. If we want to travel into the future, all we need is a machine that can move at a velocity close to the speed of light. As we approach this speed, the slower time moves. If you go fast enough, you simply get off. You will hardly have aged. Decades, or even centuries, will have passed 'back home'.

According to physicist J. Richard Gott in his 2002 book *Time Travel in Einstein's Universe*, travelling back in time is far trickier. It entails fiddling with **WORMHOLES**, cosmic strings or **BLACK HOLES**. These are the kind of time machines feasible only with mind-warping technology.

And in 2007, American scientist Ronald Mallett broke the news of his life-long struggle to build a time machine. Mallett's take on temporal travel is to bend spacetime. Massive objects such as stars and planets can do this; Mallett is among those who believe that light too can bend the continuum.

So, rather than the De Lorean envisaged in the *Back to the Future* films, Mallett's machine is a ring laser, an extremely powerful one. And maybe one day, by simply popping into this huge vortex of light, travel through time may be possible.

COUNTDOWN

Launching a **ROCKET** isn't easy. Recent space shuttle disasters have shown that things can and do go wrong. So many different actions have to occur almost simultaneously. But how can you co-ordinate things so that when the time comes everything occurs at precisely the right moment?

It was a question that science fiction director Fritz Lang grappled with when he was putting together his seminal 1929 film *Frau im Mond*, a film whose melodramatic story line of finding both gold and love on the Moon endeared it to the public at a time when the talkies were making silent films obsolete.

Lang's answer to the conundrum was so simple and neat it was immediately adopted by the burgeoning space industry. Fritz Lang, science fiction filmmaker, invented the countdown.

In the final moments before the rocket ignites, images appear on the screen (this was one of the last silent films) showing the countdown from ten to zero. It was a simple answer to a complicated problem. The speed by which it was adopted by the aeronautics business was no doubt due to the fact that the influential rocket scientist Hermann Oberth worked as an advisor on the film.

FLASH MOBS

You're sitting in a bar when suddenly you get a text message. It says simply 'Trafalgar Square midday clap for one minute then disperse'. When you arrive at the Square over a thousand others are doing the same thing. You've just taken part in a flash mob. Part performance, part social statement, it relies on two major factors – mass communication and mass transportation. The first flash mob was organised in 2003 and they are now held regularly all over the world, often for marketing purposes. Thousands of people synchronise their lives to appear at the same place, at the same time, doing the same thing.

The origins of the flash mob lie in the 1973 short story *Flash Crowd* by Larry Niven. In this tale it is ubiquitous news media that provides the communication to cover small disturbances that allow crowds to gather quickly via the mass transportation of **TELEPORTATION**. The voyeurs soon become embroiled in the conflict so that they quickly escalate into major riots, which themselves become a form of participative entertainment for the general public.

A flash mob at Somerset House in London.

Immortal: David Tennant is the tenth television Dr Who.

Non-senescence? Dr Aubrey de Grey claims that the first human to live to 1,000 years of age may have already been born.

Flash mobs have subsequently appeared in science fiction in the 1998 Bruce Sterling novel *Distraction* and even in a 2004 episode of the 'science faction' series *CSI: Las Vegas*.

IMMORTALITY

Perhaps Miss Alabama 1994 put it best: 'I would not live forever, because we should not live forever, because if we were supposed to live forever, then we would live forever, but we cannot live forever, which is why I would not live forever.' But eloquent beauty pageant contestants aside, the search for immortality has been one of man's more persistent obsessions. Although immortal figures can be found in fairytales and myths, the scientific pursuit of immortality can once again be traced to the founding mistress of science fiction. Mary Shelley's 1833 *The Mortal Immortal* features a hero who doesn't use magic to extend his life, turning instead to all that science offers. His fate is to watch those around him wither and die whilst he alone continues. He has effectively become a traveller in time, always fated to move forward.

The quest for immortality can be traced through the pages of science fiction by its various designs and science has now taken note of the plans and designs for eternal life within those pages. Shelley's work and that of others revolves around achieving immortality via biological methods, attempting to use treatments and medicines to artificially extend lifespan. These could be the results of **EUGENICS**, **GENETIC ENGINEERING**, the use of **ARTIFICIAL ORGANS** or even becoming a **CYBORG**. Science fiction exploring biological immortality has seen characters like *Khan Noonien Singh*, Captain Kirk's nemesis from the second *Star Trek* film, develop as a result. In science it is hoped that nanomedicine may hold the key. Such is the confidence of leading biogerontologist Aubrey de Grey of Cambridge University that he has established the Methuselah Mouse Prize. This monetary award will go to the researcher who can significantly extend the life of a mouse as the first step in pioneering biological immortality.

Before that happens perhaps advances in computer technology would allow us to upload our personalities to live forever on the **INTERNET**. Although Olaf Stapledon's 1930 novel *Last and First Men* comes close with its organic human-like brains grown into a machine, it is Roger Zelazny's 1967 *Lord of Light* that really explores the theme properly. In this story the crew of a space vessel achieve immortality by uploading their consciousness into new bodies so many times that they begin to see themselves as gods. Leading roboticists and cognitive science researchers working in this field, like Marvin Minsky and Hans Moravec, refer to the new consciousness

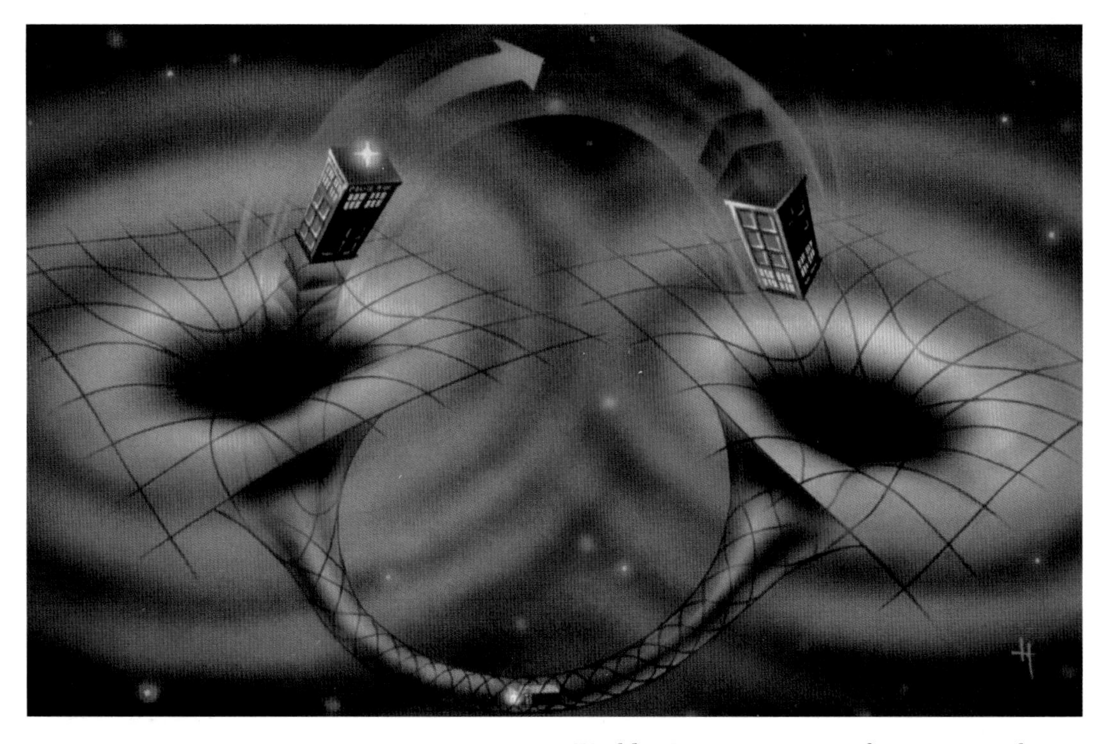

Down the wormhole: a journey through spacetime.

which uploading would form as an *infomorph* – a variety of **ARTIFICIAL INTELLIGENCE** best explored in the work of science fiction author Greg Egan. These scientists acknowledge that we do not have the processing power or technology to accomplish this task yet, although the exponential development of computing may mean that it is on the horizon.

Whilst biological and technological immortality have their place, it is genetic immortality that has made the furthest strides in science. Taking the idea that our genes codify who we are, a concept made popular by the evolutionary biologist Richard Dawkins in his book *The Selfish Gene*, its consequences are that **CLONING** becomes a method of achieving immortality. The numerous science fictional explorations of this topic are suddenly cast into a new light when we realise that identical genetic material can be made to last forever.

But perhaps we are already immortal. Theorists working in the field of quantum physics point out that according to the Many Worlds Interpretation of quantum theory pioneered by Hugh Everett, every action occurs (or doesn't) somewhere in an infinite set of universes. There therefore exists a path through your life in which you never die, every decision you ever made leads to your continued life. The chance of this occurring is phenomenally low but with an infinite number of possibilities it does exist. This is referred to as 'quantum immortality'.

WORMHOLES

You're standing at the Crucifixion.

Dumbstruck and open-mouthed, you can't help but stare at the scene. Perhaps the most famous in all of history. It was an expensive package but your time travel tour operator said it would be well worth the cash. Just a few points to remember. You must do nothing to disrupt history. (Note to self: don't tread on any butterflies this time). And when the crowd is asked who should be saved, you join in with the call, 'Give us Barabbas!'

Suddenly, you realise something about the

crowd. Not a single soul from 33 AD is present. The mob condemning Jesus to the cross is made up lock, stock and barrel of tourists from the future.

Far-fetched? Stephen Hawking thought so. The story is *Let's go to Golgotha*, a 1975 time travel tale by Garry Kilworth. Hawking suggests that the apparent absence of such tourists from the future is a strong argument against the likelihood of **TIME TRAVEL**. We may call it 'Hawking's Paradox'.

Of course, this doesn't mean time travel is physically impossible. It may merely mean that it is never developed. And even if it is developed, there may be snags.

Assume we create a **WORMHOLE**. A wormhole is a region of warped spacetime. It is essentially a 'shortcut' in space and time through which to travel. Trouble is, time travellers would not be able to travel back in time to a date before the wormhole was created. This may explain why we've not been overrun by tourists. We've simply not created a wormhole yet.

It was science fiction writer John Campbell who invented such 'space warps'. In his 1931 story *Islands of Space* Campbell used the idea as a shortcut from one region of space to another. In his *The Mightiest Machine* (1934), Campbell called this same shortcut 'hyperspace', another now-familiar coinage.

A year later, Einstein and Nathan Rosen penned a paper that nailed the notion of 'bridges' in spacetime. The imaginative American physicist John Wheeler re-dubbed these Einstein-Rosen bridges 'wormholes' in 1957.

A wormhole has at least two mouths, connected to a single throat. Such wormholes are valid solutions in relativity theory. Matter may 'travel' from one mouth to the other by passing through the wormhole. We haven't observed one yet, but the Universe is still young and we haven't been looking very long.

Of course, the idea of wormholes in the public imagination owes almost everything to fiction. Television shows such as *Stargate SG-1* and *Sliders* owe their existence to wormholes. Characters are either zipping from our Milky Way to the Pegasus galaxy, as in *Stargate*, or they are shifting between parallel universes, as in *Sliders*.

10 popular science books

- *Awakenings* (1973) by Oliver Sacks
- *The Double Helix* (1968) by James D. Watson
- *Guns, Germs & Steel: The Fates of Human Societies* (1997) by Jared Diamond
- *The Panda's Thumb: More Reflections in Natural History* (1980) by Stephen J. Gould
- *The Selfish Gene* (1976) by Richard Dawkins
- *The Seven Daughters of Eve: The Science That Reveals Our Genetic Ancestry* (2001) by Bryan Sykes
- *A Short History of Nearly Everything* (2004) by Bill Bryson
- *Silent Spring* (1962) by Rachel Carson
- *Cosmos* (1983) by Carl Sagan
- *The Ascent of Man* (1973) by Jacob Bronowski

Indeed, Carl Sagan's 1985 novel, and ensuing 1997 movie, *Contact*, created a cottage industry among theoretical physicists. Eleanor Arroway, Jodie Foster's character, travels 24 light years through an Einstein-Rosen bridge to the star Vega. Analysis of the story by Kip Thorne, as a favour to Sagan, is cited by Thorne as the initial impetus for his work on wormhole physics.

Unlike a **BLACK HOLE**, a one-way journey to oblivion, a wormhole has two mouths: an exit and an entrance. Thorne figured that some form of **ANTI-GRAVITY** might control such wormholes so they can be used as the shortcut through hyperspace envisaged by John Campbell in 1931.

With a dash of exotic matter, providing the anti-gravity pressure to keep the wormhole throat from imploding, Thorne and many other physicists believe that a stable wormhole could be created and readily turned into a **TIME MACHINE**.

Crucifixion, anyone?

TIME AS THE FOURTH DIMENSION

Your world has four dimensions. Three dimensions are space; time is the fourth. Seems obvious, doesn't it? And yet it took H.G. Wells to sort it out.

The Time Traveller in Wells' classic 1895 novel *The Time Machine* explains it simply enough: 'There is no difference between Time and any of the three dimensions of Space, except that our consciousness moves along it'.

Now, the notion was nothing new. Relating time to space had a long history, going all the way back to Aristotle. Even though industrialisation had led to a nineteenth century obsession with time, most regarded the fourth dimension as spatial. Wells begged to differ. And in so doing he was in the vanguard of a new and exciting chapter in the history of ideas. Time was in the ether.

Time splashed upon the canvas of the Cubists. Artists such as Picasso and Braque produced paintings where various viewpoints were visible in the same plane, at the same time. All dimensions were used to give the subject a greater sense of depth. It was a revolutionary new way of looking at reality.

Time was captured in cinema, and the stop-motion photography of Étienne-Jules Marey. It inspired Marcel Duchamp to paint his highly controversial *Nude Descending a Staircase*, which depicted time and motion by successive superimposed images. The Americans were scandalised.

Physics caught the fever and Einstein introduced Special Relativity in 1905. At first, he spoke of three spatial dimensions, and time. It was only after his teacher, Hermann Minkowski, promoted the view of time as the fourth dimension

❸ Chaos Theory pioneer Edward Lorenz.
❹ Chronophotograph of a man doing a high jump, taken in 1892 by Etienne-Jules Marey.
❺ Time for action – Warren Beatty as *Dick Tracy* (1990).

that the notion of the space-time continuum was created. It was essential to the development of Einstein's later work in General Relativity, and it is very precisely the concept that Wells pioneered.

The idea of spacetime was born. Einstein gave us a new perspective on the fourth dimension. Moving clocks run slow. Speeds don't add up. And time is slowed down by gravity. It was a revolution in time. And it worried Salvador Dali. His anxiety is palpable in his famous painting *The Persistence of Memory*. The floppy clocks are history's most graphic illustration of Einsteinian gravity distorting time.

The obsession with the fourth dimension continued. One fascination was the time paradox, which can be nut-shelled by the question 'what would happen if I went back in time and killed my own granddad?' The skill of tampering with time in this way reached its peak with Robert A. Heinlein's *All You Zombies* (1959). The main character of the story moves along the fourth dimension, undergoes a sex change, and becomes his/her own mother and father. A second attraction was The Butterfly Effect, an idea which first found its voice in fiction with Ray Bradbury's moral fable *A Sound of Thunder* (1952). A time-tourist wreaks temporal havoc by treading on a prehistoric butterfly and unleashing an alternative world. The story told of sensitive dependence upon initial conditions, and it was written a full ten years before early pioneer of chaos theory Edward Lorenz developed the principles for the scientific community through mathematics and meteorology.

Since the 1960s we've had String Theory. This 'Theory of Everything' suggests a more complex picture than even H.G. Wells imagined. The universe may be made up of incredibly small strings vibrating in a space-time continuum consisting of 11 dimensions: the first four effective (observable) dimensions that Wells spoke of in *The Time Machine*, plus seven minor (smaller) dimensions.

So even though we live in a four-dimensional world, there may well be other dimensions we cannot perceive. Perhaps it was on this basis that a character in Kurt Vonnegut's *Breakfast of Champions* (1973) was described as having, 'a penis eight hundred miles long and two hundred and ten miles in diameter, but practically all of it was in the fourth dimension'.

DICK TRACY'S TIME DEVICE

Time device doesn't do it justice, really. And wrist-watch certainly doesn't.

The gadget in question belonged to the lantern-jawed, big-hitting, sharp-shooting, mega-intelligent cop of the comic strip world – Dick Tracy.

The strip first dawned in the creative mind of cartoonist Chester Gould. By 4 October 1931, the strip had hit the streets, distributed by the *Chicago Tribune* syndicate.

In January 1946, the strip changed forever.

The innovation: Dick Tracy's seminal communications device. First worn as a two-way wrist radio, the gadget quickly evolved to become the strip's most desirable high-tech icon. Adorning not only Tracy, but also members of his trusty law enforcement agencies, the 'wristwatch' was an early precursor of the **MOBILE PHONE**.

A year after its first appearance in fiction, the two-way device had inspired its first imitation in fact. Weighing in at a miraculous 3 ounces, the real-life wrist radio was invented by Dr Cledo Brunetti using tiny vacuum tubes. The downside was that the radio had a range of merely a mile.

The strip moved on. Tracy's two-way wrist radio became a two-way wrist TV in 1964. By the late 1980s it was a two-way wrist PC.

In 2006 fact followed strip fiction once more with the launch of the Eurotech Zypad Wrist-Worn PC. The company is looking to retail in a variety of fields, including law enforcement.

TELEPORTATION

A classic of science fiction, teleportation – the ability to transmit matter across space instantaneously and recreate it exactly – is every schoolboy's dream. The ability is mentioned in an early Jewish myth where it is referred to as *Kefitzat Haderech* (literally 'the shortening of the way' or 'short cut'). This term was in turn altered and

The *Star Trek* transporter.
Geena Davis is trapped in a teleportation pod in *The Fly* (1986).

adopted by the science fiction writer Frank Herbert for his *Dune* series into the *'Kwisatz Haderach'*, referring to the hero of the series Paul Atriedes and the fact that he was a **GENETICALLY ENGINEERED** 'short cut' to man's future as a result of a **EUGENICS** breeding program to create a *homo superior*.

Other mythologies and magic tales also mention teleportation-like activities but these are always represented as either mystical or a divine gift. The first exploration of science being able to accomplish such a feat comes with Edward Page Mitchell's 1877 tale *The Man Without a Body*. In this short story the scientist hero, after successfully disassembling his pet cat, telegraphs its atoms and then reassembles them. Whilst trying to replicate the experiment with himself as the subject, an unfortunate power cut only transmits his head.

Used in numerous science fiction stories from that point on, **MATTER TRANSMISSION** was also an early staple of cinematic science fiction. Used in the 1939 *Buck Rogers* serial to replace stairs, teleportation quickly became a cheap way of moving characters from place to place. Nowhere is this more noticeable than in Gene Rodenberry's influential science fiction masterpiece *Star Trek*. Originally it was planned to have the characters land their ship (the *Enterprise*) on each planet but budgetary constraints on the special effects forced a more innovative solution and the transporter was born. In *Star Trek* this system usually works flawlessly. However, other science fiction has explored the potential difficulties associated with such technology.

One such potential difficulty is most graphically described in *The Fly*. Originally a 1957 short story by George Langelaan, it has been adapted into three films, all of which explore the grotesque consequences when a scientist and a fly accidentally end up fusing in a teleportation accident. Teleportation has also been fused with **TIME TRAVEL** in films like *The Terminator* and science fiction series like *Stargate SG-1* in which **BLACK HOLES** are used as **WORMHOLES** to facilitate such transit.

Most matter transmission in science fiction is described using quantum terminology. This entails the original object being destroyed and recreated elsewhere. This poses problems for some who wonder where the soul, if it exists, would go. The problem is neatly avoided if, rather than simultaneous destruction and recreation, teleportation featured the exact duplication at distance of the object or person. This does, however, raise problems as to exactly who is the 'original'.

Science has been slow to catch up with its fiction. It took until 2002 when Australian scientists successfully teleported a laser beam by scanning a specific photon, copying it and then recreating it at an arbitrary distance. The next step was for teams in both Germany and America to independently teleport ions of calcium and beryllium using very similar techniques to each other.

A further breakthrough was made in 2006 when Danish scientists successfully teleported an object. Although miniscule in scale, it was nonetheless constructed from trillions of atoms. This half-metre transmission of matter opens up an unknown future in which the science fiction visions of generations of writers look set to become a reality. No more hanging around in traffic jams – all we will need is the latest **MOBILE PHONE**. Flip it open and say, 'Beam me up Scotty'.

ALLOHISTORIES

What if you hadn't picked up this book and started to read these words? What if you had decided to eat a sandwich instead? Perhaps there is a universe in which that happened. If that is the case and someone records what might have happened, then it is generally referred to as an alternate history or allohistory. These allohistories occur in time, not space. They are distinct from **QUANTUM WORLDS** in that they are completely fictional 'what if's. These are thought experiments designed to speculate about the past.

They have their origin in Geoffroy-Château's 1836 *Napoléon et la conquête du monde, 1812–1823: histoire de la monarchie universelle* (Napoleon and the Conquest of the

➋ Napoleon is undefeated in Geoffroy-Château's 1836 alternate history.

➌ If they all land on top of each other someone's going to get hurt. Time for a time press.

World, 1812–1823: History of the Universal Monarchy) in which Napoleon is not defeated in his disastrous Russian campaign of 1812 and continues his victories across Europe.

Later tales blend **TIME TRAVEL** with these alternate histories so that the traveller is the cause of the divergence. Perhaps the most influential of these early stories is H.G. Wells' *Men like Gods* (1923) in which travellers are transported to an alternate Earth which is almost a utopia, having diverged from our own Earth history several centuries earlier. As a result of Wells these dimension-hopping alternate histories came to be incredibly popular in the early pulp magazines and on through the twentieth century. Later examples of alternate histories include science fiction series like *The Time Tunnel* (1966), *The Tomorrow People* (1973), *Quantum Leap* (1989) and *Sliders* (1995).

Social scientists turned to science fiction to provide them with inspiration for their work. The 'what if' exercises employed by scores of science fiction authors were utilised to create counterfactual scenarios in a burgeoning new field called *speculative cliometrics.*

Although historians like J.C. Squires, in his *If It Had Happened Otherwise* (1931), which is a collection of 'what if' essays by leading historians, had published counterfactual volumes, these works remained somewhat whimsical. It was left to the Nobel prize-winning academic Robert William Fogel to take a more rigorous and scientific approach. In 1964 Fogel published his study *Railroads and American Economic Growth: Essays in Econometric History*, in which he postulated a past in which America did not have access to railroads from 1890 onwards. With careful scientific analysis he demonstrated that there would have been very little difference to American expansionism if it had to rely on wagons, canals and rivers rather than the railroads. The seriousness with which other scholars took Fogel's work allowed other scientists to consider using his

counterfactual techniques to good effect within their own fields.

The academic Richard Lebow has provided a list of the use of such techniques within wider science. He includes Roger Penrose's classic *Shadows of the Mind: A Search For The Missing Science Of Consciousness* and the *Elitzur-Vaidman Bomb-Testing Problem*. The latter is a 1993 thought experiment in quantum mechanics, proposed by Avshalom Elitzur and Lev Vaidman, which uses information from non-events to test nuclear weapons. In a more concrete example in 2006, scientists at the University of Illinois ran a *Counterfactual Computation* using a quantum computer to achieve the correct answer to an algorithm without actually running it.

THE TIME PRESS

Imagine having your recovery time from an injury speeded up to a mere few seconds. In John Varley's 1983 novel *Millennium* (and its 1989 film version) scientists use a time press to do just that. In the novel, humanity is faced with extinction through depopulation. The answer is to **TIME TRAVEL** back and snatch people just before they die in car and air crashes, leaving a fake corpse behind (also explored in the 1992 Mick Jagger film *Freejack*). The victims' injuries are rapidly healed through the use of nutrients fed into the body which enhance the healing process so that scars appear in a matter of seconds.

Today's scientists are doing something akin to this (although without such dramatic effects) by using a technique called 'blood spinning'. This process extracts and spins a patient's blood in a centrifuge which concentrates the blood platelets to five times their normal level. By adding *calcium* and the enzyme *thrombin*, a gel is formed that is high in natural growth factors that accelerate healing. The gel can then be applied directly to the wound or injected into an internal injury. Introduced for trauma victims, blood spinning has become controversial because it is used throughout the sporting world to accelerate injured athletes' recovery times.

PRECOGNITION

Psi power, or psionics, sounds like something straight out of *The X-Men*.

Psionics describes a set of alleged extra-sensory powers such as telepathy (mind-reading), telekinesis (movement by the mind) and, of course, precognition (seeing into the future).

Psi power is an immensely popular motif in science fiction. The first writer to use the term psionics is probably Murray Leinster in his 1955 tale *The Psionic Mousetrap*, though Alfred Bester's *The Demolished Man* (1953) had busied itself worrying how to get away with murder in a telepathic society.

In the 1960 movie of John Wyndham's *Village of the Damned*, an alien race manage to impregnate Earthly mothers to produce psionicly powered blond-haired children who come over like a group of mutant Hitler Youth.

Most famous is the 2002 movie of Philip K. Dick's *The Minority Report* (1956), in which three 'pre-cog' mutants are used to divine **PRE-CRIME.**

Although prevalent as a theme in the science fiction of the 1950s, it wasn't until a secret visit from the CIA to the Stanford Research Institute in 1972 that researchers there started to develop techniques to assess the power of psionics, under the leadership of physicist Dr Harold E. Puthoff. Key to their strategy: remote viewing (RV) where a 'viewer' attempts to collate data via psi power on a remote target with little prior knowledge of the target's identity. Though sceptics dispute the validity of psionics, this government program ran for over two decades with some claimed success of remote viewing in the field of military operations. The existence of the program and some of its details were declassified in 1995 by President Bill Clinton in an 'open government' drive.

Village of the Damned (1960).
Putting the freshness back intelligently: the iRobot Roomba.
Instantaneous translator C3PO.

AUTOMATED VACUUM CLEANERS

Time is of the essence. In our increasingly pressured lives we have little of it left for such domestic niceties as vacuuming the house. But that's okay because we live in a science fictional world. A world in which the *Electrolux Trilobite* or *iRobot Roomba* can clean our floors for us. These automated vacuum cleaners can move through our **BUILT ENVIRONMENTS** mapping our rooms using built-in radar and methodically covering every spot on our floors. They will return to their base to recharge before moving into the fray once again. Once full they buzz engagingly to allow their bags to be emptied before scuttling off to work again.

A Cold War classic anticipated this picture of domestic comfort over fifty years ago. Ray Bradbury's 1950 short story *There Will Come Soft Rains* features **ROBOT** mice, which scurry around the room, their whiskers a-twitch for the slightest sign of dust, ensuring that all is clean before they retire to their burrows to recharge. What next? How about robot dogs that walk themselves, or even electric sheep?

INSTANTANEOUS TRANSLATORS

It's one of the staples of science fiction. The hero steps forward to exchange a few words with aliens who speak a strange language. Unable to communicate, he uses an instantaneous translator to, er, instantly translate the language, thus avoiding any difficult misunderstandings. Examples include the universal translator from *Star Trek*, C3PO from *Star Wars*, the babel fish from *Hitchhikers* and the TARDIS (which also travels in time) from *Doctor Who*. Always based on the same premise, the idea goes that the device, computer or **ROBOT** has a massive store of languages in its databanks, often including dead or dying languages. C3PO can translate over six million forms of communication.

Whilst not up to that standard, the Rosetta Project, which started in 2000, is a digital store held online which has currently archived

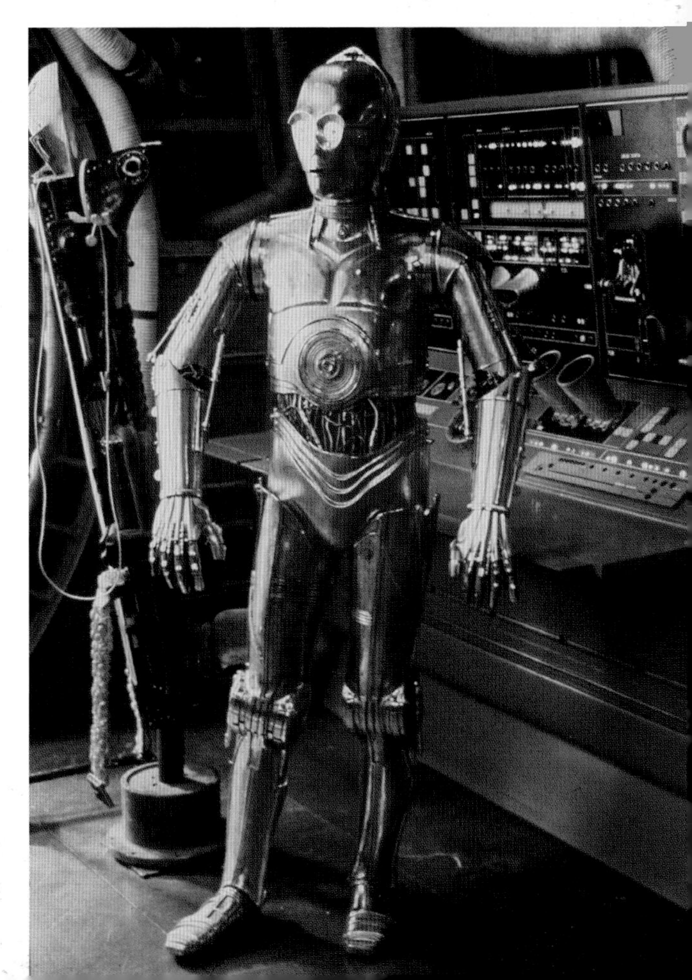

over two and a half thousand of the world's languages, with more being added all the time. It has also produced the Rosetta Disk. This three-inch nickel disk is etched with 27 thousand pages of nano-sized text. This allows the user to access all the languages merely through magnification. It is hoped it will last at least ten thousand years or until the **END OF THE WORLD**. Whichever comes first.

TIMEBROKERS

Ever had a couple of hours to kill and thought to yourself, 'I wish that I could see my favourite celebrity'? Perhaps a timebroker can make it happen for you. In our 24/7 world people want entertainment at all hours of the day and night and it's the timebroker's job to arrange it for them. Paul Di Filippo coined the idea in 2005 with his science fiction short story *Shuteye for the Timebroker.* Set in a world where people were always awake, the entertainment industry had to adapt and timebrokers were the answer.

It didn't take long for the **INTERNET** to respond. The social networking site *eventful.com* launched a new feature called *eventful demand* in 2006. In it users could request events that hadn't happened yet. They suggested dates, venues and times when they were free. If enough people signed up to the petition, the website then got in contact with the artists and passed on the request for the event.

The first successful event to be brokered took place on 2 July 2006. It has snowballed since then. Top-selling American R & B group Pretty Ricky announced a 2007 tour entirely generated by users and managed by the timebrokers at *eventful.*

GUIDED EVOLUTION

'A scientific definition of God.' That's how Stanley Kubrick described his 1968 movie masterpiece *2001: A Space Odyssey.*

Darwin had stirred German thinker Friedrich Nietzsche to recognise three phases in the evolution of humans: ape, man and, finally, superman. As Nietzsche said, 'What is the ape to man? A laughingstock, or painful embarrassment. And man shall be to the superman: a laughingstock or a painful embarrassment.' Modern humans are solely the stepping-stone from ape to superman.

The trouble is, neither Darwin nor Nietzsche is exactly primetime TV fodder. And there is little drama in Darwinian evolution. Just the sluggish, unsolicited course of creeping change. So, along with famous science fiction author Arthur C. Clarke, Kubrick souped-up Nietzsche. And they souped-up Darwin: they developed a fictional form of Stephen J. Gould's 'punctuated equilibrium'. In *2001*, dumb, blind evolution is ruptured by

◀ The pretty boys of Pretty Ricky, available for gigs on demand.
▶ Humanity gets a leg-up onto the evolutionary ladder in *2001: A Space Odyssey* (1968).

the periodic guiding hand of an elusive **ALIEN** race. It is a story of the effective creation and resurrection of man in three easy parts. A scientific definition of God, in effect.

As the movie's title implies, the story takes us on a journey through time and space. We start with the subhuman ape, and end up with the post-human starchild. *2001* is a four-million-year filmic story, whose unfolding embraces each theme of science fiction: space (alien contact through the monoliths), time (evolutionary fable), machine (HAL, the computer turned murderer) and monster (human metamorphosis).

The odyssey begins with a small band of apes on the long, pathetic road to racial extinction. But they are saved. And their salvation comes in the form of guided evolution. The mysterious presence of the black monolith transforms the hominid horizon. The journey to superman begins.

In a single frame of film, the space age dawns. It is a bland future, dominated by corporations and technology. Ironically, the most 'human' character is the robustly intelligent spaceship computer, HAL 9000. The potent evolutionary force imparted by the black obelisks is once more overdue. The space age was ultimately inspired out of the apes by alien intelligence. And so it will be with man.

Under the watchful presence of the monoliths, modern man, in the form of astronaut David Bowman, comes to an end. With the immense presence of planet Earth filling the screen, the foetus of the superhuman starchild glides into view. Moving through space without artifice, the image suggests a new power. Man has transcended all Earthly limitations.

Three was also the magic number for Béla Bánáthy: the famous Hungarian linguist and systems scientist was inspired to identify three seminal events in the evolutionary journey of our species.

In his book, *Guided Evolution of Society*, Bánáthy suggests the first decisive episode occurred some seven million years ago, when our hominid ancestors first developed. The second crucial event happened when the *homo sapiens* began the revolutionary route of cultural evolution. And today, says Bánáthy, we have arrived at the third major event: the revolution of conscious evolution. We now have the power and responsibility to guide the evolutionary journey of our species.

CRYONICS

We can travel through time. Some of us have already travelled through time. At least that's what it felt like. We accomplished this task not with a **TIME MACHINE** but with a fridge, through the science of **CRYONICS**. Clinical medicine is now able to switch people off for up to an hour leaving them with no heartbeat or brain activity. Although used for certain surgical procedures, the effect on that person is as if time stopped and restarted an hour later. Further breakthroughs are expected with the development of vitrification technology. This technology allows for organs to be frozen and then successfully defrosted and used. In 2005 a rabbit's kidney was successfully re-implanted after being frozen at −135°C.

The idea of using a deep freeze to travel through time was explored by Jack London's first published work, *A Thousand Deaths* (1889). In this the hero is repeatedly killed and brought back to life by his mad scientist father: 'after being suffocated, he kept me in cold storage for three months, not permitting me to freeze or decay. This was without my knowledge, and I was in a great fright on discovering the lapse of time'. From that point on extreme cold has been used throughout science fiction to help heroes and villains travel through time.

MODULAR HOUSING

Houses just keep getting more and more expensive. We can't build our houses quickly enough. There just isn't the time to

➊ Frozen in time: *Austin Powers* (1997).
➋ The Spacebox is a $22m^2$ studio apartment designed by Dutch architect Mart de Jong that can be erected in days.
➌ The Doomsday Clock, symbol of the *Bulletin of Atomic Scientists*.

erect all of the dwellings needed to supply demand. Perhaps the answer is modular housing. Bolt together units that can be tailored to individual specifications.

In his 1992 cyberpunk classic *Snowcrash* Neal Stephenson (who conceived both the **AVATAR** and **VIRTUAL GLOBE**) came up with the answer. The lower echelons of his population could be found eking out an existence in shipping containers. That notion has now been put into practice by the visionary architect Adam Kalkin, who has designed and produced the Quik House, a home constructed from these ubiquitous containers. Retailing for around

£85,000, this three-bedroom, fully-fitted house takes only twelve weeks to put together. With stainless steel appliances, a working fireplace and wall-to-wall carpeting, as well as two and a half bathrooms (one en-suite), it's not exactly a slum. Erected in the time it takes your typical builder to sink the foundations of a reasonably sized conservatory, the Quik House is the speedy and cheap answer to all your housing needs.

THE DOOMSDAY CLOCK

Doomsday isn't what it used to be.

The days of yore imagined the end of the world as pure fantasy: four horsemen of the apocalypse riding out of chapter six of the Book of Revelations. All that changed with the rise of modern science. Fiction was now able to imagine more technologically plausible ways of destroying the world.

Mary Shelley's lesser-known classic *The Last Man* (1826) featured a future industrial world stricken by a global plague. M.P. Shiel's

⬆ FTL travel by spaceship…
➡ …and FTL travel by police box.

timeless post-apocalyptic tale *The Purple Cloud* (1901) saw the calamitous leak of a chemical gas, killing all the people on the planet.

Then H.G. Wells 'invented' the **ATOM BOMB.** His 1914 novel *The World Set Free* was read by Hungarian physicist Leo Szilárd in 1932. Szilárd went on to actually build such a bomb a year later.

The fictional doomsday device became fact. The threat of global nuclear war loomed, and the Doomsday Clock became its symbol.

Since 1947 the symbol's clocksmith has been the *Bulletin of the Atomic Scientists* at the University of Chicago. It uses the analogy of the human race being at a time that is 'minutes to midnight'. Midnight is doomsday, the end of time.

The clock currently reads 11.55. Thanks to H.G. Wells.

FASTER-THAN-LIGHT TRAVEL

Consider this: Albert Einstein is flying steadily through the cosmos at the speed of light.

In front of Albert, and at arm's length, is a mirror. The mirror is also travelling at the speed of light. The question is, if Albert looks into the mirror, does he see his own reflection? At the tender age of sixteen (in the same year H.G. Wells wrote *The Time Machine*), this *Gedankenexperiment*, or thought experiment, puzzled Einstein greatly.

Since Albert is effectively sitting on top of a light beam, light from him would never catch up with the mirror. His image would disappear. This all struck Einstein as rather odd and he did not believe it. His solution to this puzzle was revolutionary: re-write physics and the concept of time.

Before Einstein, the maximum speed possible was thought to be boundless. But he proposed that everyone sees the same speed of light no matter how quickly they are moving. Not only that, Einstein showed that the speed of light was the maximum speed.

Einstein then wondered what happens if you try to get a massive object (we might propose the starship *Enterprise*) to go beyond the speed of light. This light-speed barrier is one of the results of Special Relativity, developed by Einstein in 1905.

To speed along, you need energy. To travel at the speed of light, the amount of energy needed to propel you swells to infinity. To move the starship *Enterprise* at the speed of light would take all the energy in the Universe, in fact. Just a small snag.

Science fiction, of course, is also a kind of thought experiment. The superluminal speculation of fiction began with French astronomer and writer Camille Flammarion. Flammarion wrote a pre-Einsteinian fantasy, *Lumen* (1867), in which some of the 'relativistic' effects of faster-than-light travel are predicted.

Flammarion invented spacetime. Thirty years before Einstein, *Lumen* was the earliest novel to propose that time and space were not absolute. They exist, said Flammarion, only relative to one another. He also explained how travelling faster than light would render history in reverse.

10 cool spaceships

- Battlestar Galactica – *Battlestar Galactica* (1978)
- Millenium Falcon – *Star Wars Episode IV: A New Hope* (1977)
- Serenity – *Firefly* (2002)
- Nostromo – *Alien* (1979)
- Shadow Battlecrab – *Babylon 5* (1994)
- TAC fighter – *Starship Troopers* (1997)
- Tardis – *Doctor Who* (1963)
- USS Enterprise – *Star Trek* (1966)
- USS Voyager – *Star Trek: Voyager* (1995)
- X-Wing fighter – *Star Wars Episode IV: A New Hope* (1977)

Much of the early pulp fiction of the twentieth century chose to ignore Einstein's findings. Soon enough, though, such ignorance became unfashionable. In its place popped up a cottage industry of ideas and devices – the 'space warp' into 'hyperspace', **BLACK HOLE** and **WORMHOLE** travel, and more recently the tachyon drive.

Tachyons appear in Gregory Benford's *Timescape* (1980) as a means of faster-than-light communication. Such sub-atomic particles can never slow down or stop moving. They can decelerate to the speed of light, but never less than that. Could be handy. Trouble is, they're hypothetical.

In the imagined world of *Star Trek*, the warp drive is the preferred form of faster-than-light propulsion. So sophisticated is the warp, indeed, that spacecraft jauntily zoom to many multiples of the speed of light.

Warp drive fever has caught on. It has infected videogames such as *Stars!* and *StarCraft*, and fictional universes such as those in the movie *Starship Troopers* (1997) and the television programme *Red Dwarf* (1988–1999).

Mexican theoretical physicist Miguel

Alcubierre got the bug too. He theorised a type of warp drive. The Alcubierre Drive would involve a 'warp bubble' enclosing a spaceship. Space at the front of the ship's bubble would swiftly contract, while space at the rear swiftly expands. The result? The bubble would reach a distant destination far faster than a light beam moving outside the bubble.

TIME SHIFTING

In the bustling world of twenty-first century consumer fetishism, everything is on demand.

Movies, music, TV, podcasts, or just plain old news, you can download your personally selected diet of digital data to satisfy your need for instant gratification and ease of use.

That's time shifting: the recording of programmes to a storage medium for future sampling at a time more convenient to the consumer.

Pure twenty-first century, yes? Think again. Think 1889, in fact. Because that's when Jules Verne first conjured up the idea. Verne's story *In the Year 2889* made this startling prediction a full twenty years before the first news broadcasts hit the ether.

In Verne's tale, users are able to record the news and listen to it later: 'Instead of being printed, *The Earth Chronicle* is every morning spoken to subscribers, who, from interesting conversations with reporters, statesmen and scientists, learn the news of the day.'

These essential 'podcasts' are recorded by a 'phonograph' and time-shifted so that subscribers may mull when they are in a mood to do so.

↑ Grab yourself a TiVo to start time shifting.
→ The world in entropic decay...

ENTROPY

Picture yourself on a beach. The last beach on Earth in fact. You had set the controls of your time machine to the max. Now you find yourself down the foggy ruins of time. You've journeyed to the far future.

The Earth is locked by tidal forces. The planets spiral towards a red giant Sun, which hangs motionless in an endless sunset. The solar system is in meltdown. So ends H.G. Wells' terrible account of the end of the world in his famous 1895 novel *The Time Machine*.

Wells took a momentous leap in the portrayal of Darwinian evolution. He depicted the entire Universe as a machine, running down on energy. The cosmos is in a state of entropic decay, and man is being swept away 'into the darkness from which his universe arose.'

With the development of cosmology as a science in the 1960s, astronomers realised Wells may be right. This image of dying planets, dying suns and a drift in entropy became known as the 'Heat Death of the Universe'. Indeed, heat death remains a possible final outcome, with no free energy and a future in a cold, dark, empty space. The Universe will have reached maximum entropy.

Happy days.

THE END OF THE WORLD

The end of the world is coming. Just how near we are to it is anybody's guess. However, scientific concern about the coming apocalypse has been largely driven by science fiction.

Whilst entropic religious imagery appears in many faiths, it took science fiction to start the ball rolling with sustained exploration of how our pale blue dot might finally finish. Mary Shelley sat writing what some argue is the first work of science fiction during 1816, the so called 'year without a summer', in which the skies were darkened by volcanic ash following the eruption of Mount Tambora.

Shelley's second work of science fiction directly follows the apocalyptic context established by the Tambora explosion. *The Last Man* (1826) deals with the effects of a worldwide plague, which drifts across the face of the globe leaving a single survivor.

H.G. Wells was profoundly influenced by Shelley's work and it was his vision of the explosive devastation caused by an **ATOMIC BOMB** which was to be the defining focus of scientific concern about the end of the world in the twentieth century. It was not the only explosive concern, however. The 1908 Tunguska event in which a cometary fragment airburst some five miles above Siberia resulted in an explosion one thousand times more powerful than that of the **ATOMIC BOMB** detonated at Hiroshima. Science has responded to science fiction's warnings with the *Spaceguard* scheme, a global network of facilities which watch the skies for future collisions.

Shelley's vision of a plague decimating mankind was brought to life in the 1918–19 Spanish Flu pandemic in which it is estimated that five percent of the world's population died in a little over twelve months. This has led modern science fiction films like *The Andromeda Strain* (1971) and *28 Days Later* (2002) to speculate on future pandemic apocalypses such as one possibly caused by Bird Flu. The World Health Organisation takes the risk so seriously that it now co-ordinates responses to such outbreaks across the world.

Yet it is visions of atomic apocalypse which drove many science fiction creators and scientists forward. The detonations in Japan in 1945 and the ensuing Cold War with its test explosions saw a sequence of science fiction texts warning of impending doom, such as Nevil Shute's relentlessly morbid *On the Beach* or the madcap yet perhaps more powerful *Dr Strangelove*, both of which explored the circumstances and effects of a nuclear war. The plots of the thrillers *The China Syndrome* (1979) and *Silkwood* (1983) both explored cover-ups at nuclear power stations and spoke of the risks associated with running such facilities. Risks which were born out in the 1986 Chernobyl Disaster when reactor number four exploded, sending a plume of radioactive debris drifting across the Soviet Union and Europe. This event and popular coverage of nuclear disasters in fiction contributed to the tightening up of safety procedures in power stations across the world.

The overwhelming apocalyptic scenario faced by humanity in the twenty-first century would appear to be climate change. Explored by J.G. Ballard in his 1962 novel *The Drowned World*, in which global warming has raised sea levels, it has now become a staple for both fictional explorations (*The Day after Tomorrow* (2004)) and scientific warnings (*An Inconvenient Truth* (2006)). Still, if we do mess up this planet we can always travel into space, find another one and try again.

↓ *Outbreak* (1995) focuses on a pandemic caused by an ebola-like virus.

→ The 1908 Tunguska event was one thousand times more powerful than the Hiroshima atom bomb.

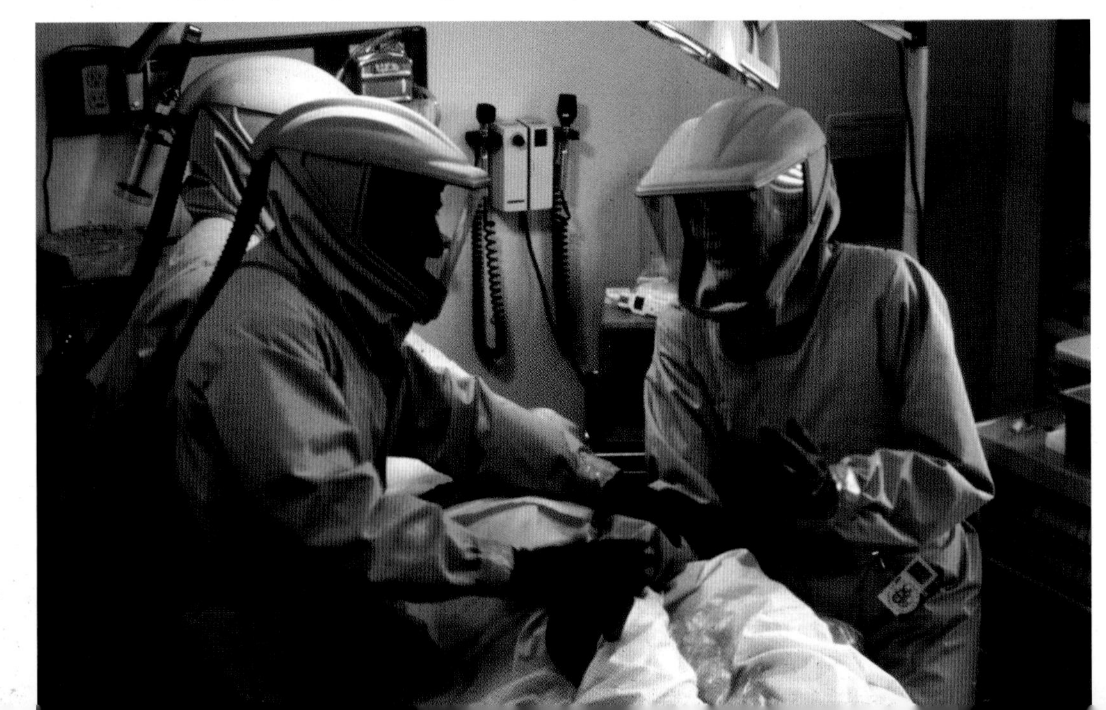

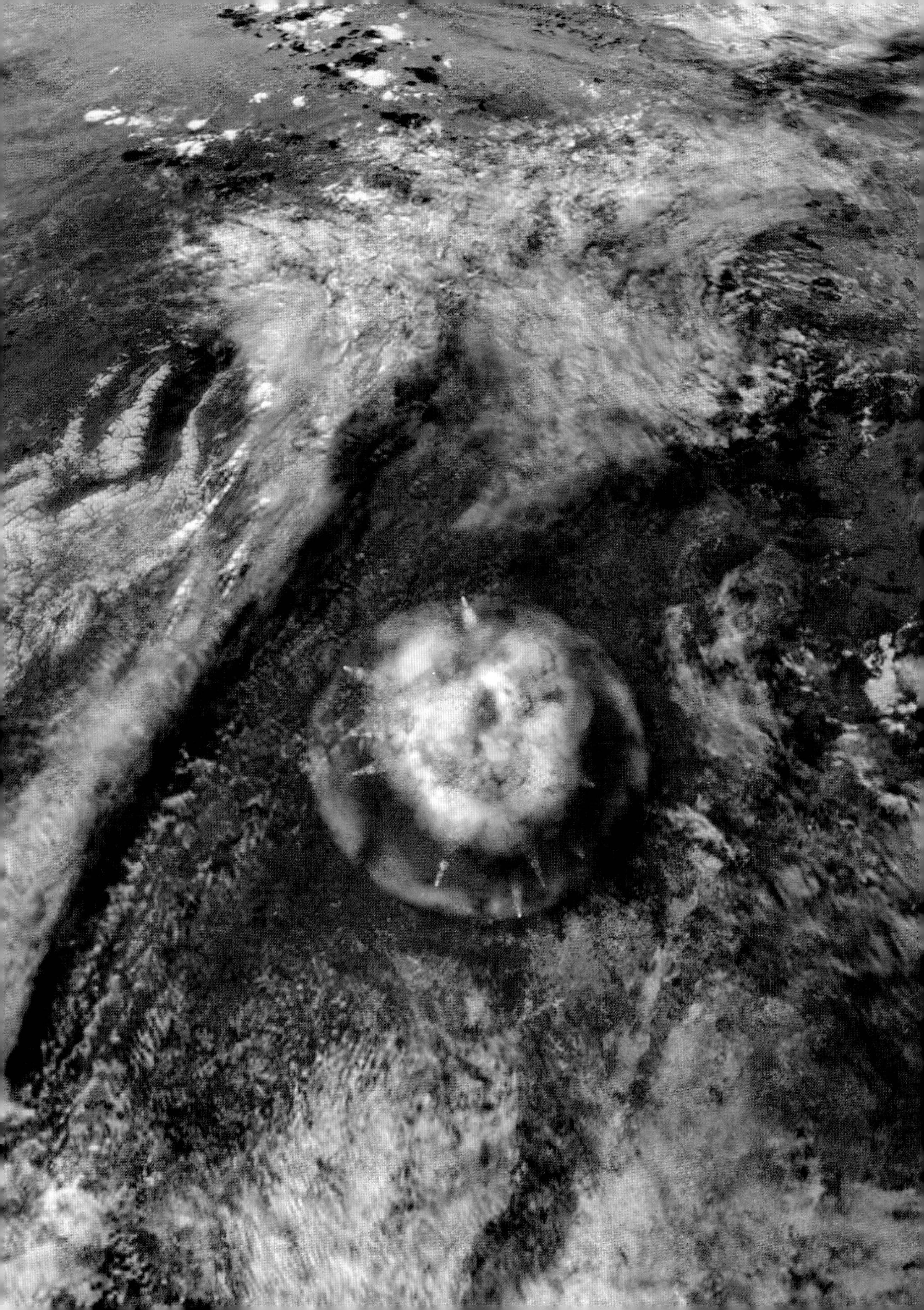

MACHINE

How do we create devices which can serve humanity without sacrificing some of vhat it means to be human? When do man ınd machine achieve a symbiosis so that :hey become a new form of life? Is technol- ɔgy neutral or can some machines truly be lescribed as evil?

Contemporary science fiction, like the con- :emporary society it is created for, is often ıpprehensive of machines. These days, the gizmos ve see trundling across the silver screen rarely seem designed to tuck us into our beds at night. nstead, a legion of Terminators, Replicants and \Is seem intent on disembowelling us and wear- ng our entrails as a hat. They seem to become ɔarticularly piqued when we trespass on their :erritory. Thus when we jack into many of the ʻirtual worlds depicted in science fiction, there ılways seems to be some psychotic machine waiting for us, hellbent on mechanical mayhem ınd trying to melt our synapses.

One of the most famous machines explored n science fiction is the robot. The hoary old :hestnut of the mad inventor manufacturing a ıumanoid automaton only for it to turn on its :reator could not offer us a clearer message: :he creation of a new technology often has ıegative consequences.

Yet this was not always the case. Science fic- :ion's past is littered with imaginative attempts :o envision a future in which machines are ɔur friends, utopian visions of gleaming netal spires in which labour-saving legions :oil industriously to serve our every whim. Perhaps we have become too mistrustful of :he machine. In a world where we rely upon nachines in such areas as transportation, nedication and entertainment, why do we still ʻiew every device as a potential handmaiden ɔf destruction?

However, there is no more striking example ɔf the exuberance with which science fiction ɔnce embraced the machine than the space- :hip. It is this device more than any other that reminds us that at heart we are inventors anc explorers, constantly seeking to push bounda- ries. It is no coincidence that the opening narrative of one of the most influential sci- ence fiction products ever created was basec on the words of Dr James Killian, science advi- sor to President Eisenhower, who remarked ir the proposal for a national space program tha 'It is useful to distinguish among four factors which give importance, urgency, and inevita- bility to the advancement of space technology The first of these factors is the compelling urge of man to explore and to discover, the thrust o curiosity that leads men to try to go where nc one has gone before.'

In turn, acknowledging their debt to *Star Trek* for helping to popularise humanity's jour- ney into space, it was perhaps inevitable tha NASA in 1976 should bow to pressure (in the form of a write-in campaign from fans of the show) and change the name of its first space shuttle from *Constitution* to *Enterprise*.

Within these pages you will find examples of some of the concepts, principles, technolo- gies and machines which have crossed from fact to fiction and from fiction to fact.

To claim that a particular scientist o author is the 'inventor' of a technology is tc dismiss the armies of scientists and writers and engineers and artists who have contrib- uted towards making the initial spark a reality

Mixed in with 'big picture' concepts like jacking in, the internet, cyberspace and the robot, we also find more mundane entries like the joystick, screensavers and the mobile phone. All these and more have con- tributed to the creation of the futureworlc that we inhabit. From the way we are able to intstantly communicate with the world, tc the ability we have to live our lives withou ever having to talk to anyone at all, science and science fiction have been busy inventing the future and, as you will see, the machine has played a pivotal role.

← The ATLAS detector (a toroidal LHC apparatus) is a giant machine designed to measure particles that emerge from the Large Hadron Collider (LHC) particle accelerator at CERN, Geneva.
↘ Metal Maria from Fritz Lang's *Metropolis* (1927).
→ Honda's humanoid robot P3.

ROBOTS

Mankind has always been fascinated with the idea of creating life. Mary Shelley deals with exactly that premise in perhaps the most famous science fiction story of all, *Frankenstein* (first published anonymously in 1818). However, a more common motif within this area has been the creation of mechanical life. Greek mythology is full of attempts to make robots. Not only did Pygmalion the sculptor have a statue he fell in love with brought to life, but Daedalus used quicksilver to give voices to his statues and Hephaestus the smith created an artificial man of bronze named Talos.

By the middle ages a variety of automata had been created which mimicked both animal and human forms. Indeed Leonardo da Vinci used the anatomical research demonstrated in his famous drawing *The Vitruvian Man* to design a mechanical knight with some limited movement. However, it was in the work of Czech science fiction author Karel Capek that these different inventions were to find their label. His 1920 work *Russum's Universal Robots* is the first documented use of the word. Drawn from the Czech by Capek's brother Josef (Karel wanted to use *labora* from the Latin), in Capek's story his robots are of biological design, although constructed – not bred. They live a life of servitude and drudgery hence the derivation from the Czech word 'robota', meaning 'forced labour'.

Very quickly, devices which were hitherto referred to as automatons of various kinds were named as robots, with films like the 1927 *Metropolis* with its depiction of Metal Maria and serials like the 1936 *Flash Gordon* showing Ming the Merciless's 'annihilants' helping to popularise the form. It was, however, in literature that robots really came to life, particularly in the work of science fiction author Isaac Asimov who famously conceived his *Three Laws of 'Robotics'* (a term which Asimov himself coined). These laws found their form in his 1942 tale *Runaround* and provide a set of hierarchical rules which govern the actions of robots and ensure that humans are not harmed. Handy then since the first real life casualty of a 'robot' was in 1979 when Robert Williams of Michigan had his head crushed by a malfunctioning industrial robot designed to sort and retrieve car parts. Since that time accidents have escalated until now there are on average over seventy incidents involving robots a year in the UK alone.

The difficulty comes when science tries to define what a robot is. Science fiction provides us with the model but science can't quite make up its mind. Japan, for example, has a much broader definition of what a robot is than that supplied by the International Standards Organisation's ISO 8373. For Japan much automated machinery is considered robotic. However, none can argue that newer household models like Asimo (surely named, in part, after the author of the *Three Laws*), or the domestic robot Wakamaru are finally living up to the science fictional visions.

Perhaps this is why people are starting to take the issue of 'Robot Ethics' seriously and trying to put rules in place to stop future misunderstandings and accidents. This is why the South Korean Government has announced it is consulting on the drafting of a Robots Ethics Charter to govern and control the development and implementation of robots in the future. The South Korean Government's goal that all of its households should have domestic robots by 2020 means that the future could bring man and his new best friend into direct conflict. With proper rules and regulations the nightmare vision of a futureworld overrun by mechanical mayhem should not come to pass…We hope.

FLYING CARS

Some people are never happy. The future hasn't really arrived, they say. We should be living in a world of silver flame-retardant jump suits, ray guns and X-ray specs.

Some go even further. They believe the time has come to hold science fiction to task. This may be a world of impressive technologies but, dude, where's the flying car?

Well, the flying car has been with us for some time. The earliest story equipped with a flying car was probably Jules Verne's *Master of the World* (1904). Verne's car not only flew; it could also double as a boat or submarine.

By 1928, Henry Ford had realised the concept in fact. But the first attempts with Ford's 'sky flivver' were troubled; a pilot died in an early test flight. In 1956, cruise missile engineer Moulton Taylor unveiled the 'Aerocar'. Cruising at up to 100mph, the little yellow car proved far more impractical than its fictional counterparts subsequently seen in the popular 1960s TV cartoon *The Jetsons* and featured in films such as *Blade Runner* (1982) and *The Fifth Element* (1997).

The latest model is Paul Moller's futuristic 'Skycar M400'. With a helicopter take-off, smooth flight, and comfortable drive, the Moller 'Skycar' is almost Vernean in its completeness. The downside: in this world of sustainability, fuel costs and air traffic control, we're still some way off those mesmerising skylines full of flivvers.

Flying cars in Luc Besson's visually stunning *The Fifth Element* (1997).

The Moller Skycar M400.

Recon Optical's Common Remotely Operated Weapon System.

At the beginning of 2007 it was estimated that there were 2.7 billion mobile phone users worldwide.

Remote Control Warfare

Warfare is becoming more and more hazardous. We seem to be improving our ability to maim and kill each other with increasing efficiency. So when in 2004 the US Army deployed a weapons system which was operated by remote control it was praised for removing its soldiers from harm's way. The CROWS (Common Remote Operated Weapons System) can be attached to a wide range of vehicles including the Hummer. It allows its operator to control a variety of weapons via a joystick from the inside of the vehicle whilst watching events on a camera and television system. What is disappointing is that it took one hundred and one years for the vision to become a reality. H.G. Wells' 1903 short story *The Land Ironclads*, first published in *The Strand Magazine*, features one-hundred-foot-long tank-like machines which deploy remote control rifles operated by joysticks to protect the soldiers inside from the perils of the battlefield. Instead of using television, Wells imagined 'sights which threw a bright little camera-obscura picture into the light-tight box in which the rifleman sat below'.

Mobile Phones

You want to call a friend so you reach into your pocket and pull out the familiar clamshell shape. Flipping it open, you dial and soon hear your friend's voice reply from thousands of miles away. Certainly a common occurrence in our modern world due to the ubiquity of the mobile phone – but in 1638? It was in that year the Bishop Francis Godwin's posthumous work of science fiction *The Man in the Moone* was published. In it the hero uses a variety of methods to communicate with his servant, including wireless sound messages, whilst marooned on St Helena. Later expanded upon in his 1657 *Nuncius Inanimatus* as a wireless telegraph system, it was the forerunner of today's mobile communication systems. In reality, it would take another three hundred years before mobile telephony was to emerge from the wireless communications pioneered by two-way radios and ship-to-shore telephones. It was only on 3 April 1973

when Motorola employee Dr Martin Cooper decided to make a call to his rival Joel Engel (the head of research at AT&T's Bell Labs) while walking the streets of New York City that the era of the mobile phone truly began. Cooper states that his concept design was inspired by the famous communicators on *Star Trek*, enabling us all to 'boldly call where no-one has called before'.

THE LIGHT SABRE

Surely it's the coolest weapon of all time? Scottish actor Ewan McGregor thought so. In eager anticipation of playing Obi-Wan Kenobi in the *Star Wars* prequel trilogy, McGregor confessed to practising his light sabre technique whilst making appropriately bristling, energetic laser noises himself.

The light sabre is associated with the Jedi Knights and is an elegant weapon usually consisting of a cylindrical hilt, a blade formed from a tight loop of highly focused light, about one metre long, which upon activation emits a coloured blade of pure energy. It is essentially a laser of immense power. And it is able to penetrate and cut most solid materials with little resistance, except for another light sabre blade, of course.

One of the closest things to a light sabre that science has come up with is the plasma gun. Plasma guns are used to coat surfaces with thin films, and in the construction of aircraft engines. They work by blasting large volumes of gas through an electric arc. This energises the gas to a plasma state, dividing it into positively charged ions and loose electrons. Plasma radiates brilliant light and heat.

But the electrons continue to be attracted to the ions. They re-attach themselves as soon as the energy diminishes, which makes plasma unstable. Sadly that makes it tricky, at the moment, to make a Jedi light sabre.

MACHINE INTELLIGENCE

'Are you serious? – do you really believe that a machine thinks?' wrote Ambrose Pierce in the opening line of his 1909 classic short story *Moxon's Master*. His reply was that man was a biological machine and as it could

think, could not a mechanical machine do the same? Machine intelligence is a subject that was explored by science fiction long before science got around to it.

Although most commonly focused on **ROBOTS** it was the scientific exploration of calculating machines which started the ball rolling. Blaise Pascal devised a rudimentary calculator in 1642 when he was only nineteen; it only added and subtracted. This was extended by Gottfried Liebniz in 1671 who added multiplication and division to its functions. It was the British inventor Charles Babbage, though, whose pioneering work on programmable computing machines in the 1820s laid the basis for science fiction. Up to that point science could only conceive of such innovations as arithmetic tools.

However, it took the British novelist Samuel Butler to extrapolate Darwin's theories of evolution into the world of machines. His 1872 utopian novel *Erewhon* (an anagram of 'nowhere') dealt with a hero who travelled to a fictional **LOST WORLD**. There he found a society which banned technological evolution beyond its most basic level. Their fear was that the machine could evolve and become intelligent, eventually enslaving its human masters: 'Complex now, but how much simpler and more intelligibly organised may it not become in another hundred thousand years? or in twenty thousand? For man at present believes that his interest lies in that direction; he spends an incalculable amount of labour and time and thought in making machines breed always better and better; he has already succeeded in effecting much that at one time appeared impossible, and there seem no limits to the results of accumulated improvements if they are allowed to descend with modification from generation to generation'. The perils of allowing machines to

The Light sabre, the coolest weapon of all time.
'Programming' the ENIAC calculator by changing the wiring between its 18,000 valves.
Arnold Schwarzenegger as highly intelligent machine in *The Terminator* (1984).

The liquid metal T-1000: molded as easily as memory.
The iconic built environment from *Metropolis* (1927).

think were therefore first questioned in science fiction before they were even depicted in the genre. It was left to the later generation, including Ammianus Marcellinus with his 1927 short story *The Thought Machine*, and John W. Campbell with his 1935 *The Machine*, to describe thinking machines in detail.

It was with the post-war creation of the *ENIAC* (Electronic Numerical Integrator And Computer) in 1946 that science started to drive forward these science fiction visions of machine intelligence. The *ENIAC* was the first large-scale, electronic, digital computer able to be reprogrammed to decipher a wide variety of problems. In this case it was artillery firing tables for the US Army's Ballistic Research Laboratory. Already the thinking machine was associated with violence and destruction, a theme picked up by Stanley Kubrick and Arthur C. Clarke in their joint 1968 film project *2001: A Space Odyssey*. In the middle sequence of the film it is the spaceship *Discovery's* onboard machine intelligence HAL 9000 which, having gone insane as a result of a programming contradiction, attempts to kill the crew. The menacing actions of HAL are belied by his lack of a body and soft voice. Similarly Skynet, the machine intelligence in the 1984 low budget classic *The Terminator,* has to find a way to 'reach out and touch' the soft squishy humans it wishes to eliminate. It does this by utilising a wide variety of autonomous killing machines to do its will. In reality computer science is still some way behind.

Despite the fact that at the 1956 conference in which the field of artificial intelligence was founded, researchers believed that the creation of a machine with intelligence equivalent to that of a human was only a few decades away, developments have failed to live up to the science fictional vision. Whilst the billions spent have helped make significant inroads into the field and helped to develop machine translation, optical character recognition, industrial robotics, speech recognition, data mining and search engines, the complete package remains elusive... at least for now.

LIQUID METAL

Imagine the benefits of being a Terminator T-1000, the fictional android assassin. It's cool enough being a shape-shifter. You easily absorb collateral damage, melt through narrow gaps, steal through prison bars. Fashion your hands and arms into any

weapon you wish to yield.

The liquid metal menace from James Cameron's 1991 movie *Terminator 2: Judgment Day* was not science fiction's first foray into fluid alloys. Abraham Merritt's *The Metal Monster* (1920) showed similar metallic potential, as did Jack Williamson's 1928 story *The Metal Man*.

The melting point of most metals is in thousands of degrees. Some, such as mercury, are liquid at room temperature. One class of metals – shape memory alloys – contract when heated up and return to their initial shape on cooling. Scientists say they may soon use such alloys as muscles in **ROBOTS**.

But the material most like the liquid metal of the Terminator series is metallic glass. Since its atomic structure is rather haphazard, it is springier than steel, more malleable than most metals, and can be molded as easily as memory.

BUILT ENVIRONMENTS

Darwin was a poet. No, not young Charles, but his ingenious grandfather, Erasmus Darwin.

Erasmus was a brilliant mechanic, inventing a speaking machine, a mechanical ferry and a **ROCKET** motor long before the dreams of Russian rocket pioneer Tsiolkovsky. But he was also a celebrated poet. Indeed, the Romantic poet Samuel Taylor Coleridge described Erasmus as 'the first literary character of Europe, and the most original-minded man'.

You can see why. One of Darwin's poems, *The Temple of Nature*, published one year after his death in 1802, enjoyed a supreme science fictional vision. It foresaw, with unerring accuracy, an overpopulated future of cars, nuclear submarines and colossal skyscraper cities.

Five years earlier the 'grandfather of skyscrapers' had been built. The oldest iron-framed building in the world, the Flaxmill in Shrewsbury, England, was built in 1797, with a fireproof combination of cast iron columns and beams; it prefigured more recognisably modern skyscrapers like the ten-story Home Insurance Building in Chicago built in 1884–85.

Their growth during the Industrial Revolution made cities even more of a focal point of civilisation. Images of the future city in fiction became iconic. In writing and in art, the city developed as a recurrent image of our hopes and fears for the future.

Fritz Lang's 1927 science fiction film *Metropolis* is a case in point. The movie highlighted disenchantment with metropolitan life and was also a seminal influence on design ideas for the modern built environment.

Produced in Germany during the height of the Weimar Republic, it was the most expensive silent film of the day, costing around 7 million Reichsmark (about $200 million in today's money). *Metropolis* is set in a futuristic urban dystopia and examines the perceived social class crisis of capital and labour.

Architects Le Corbusier and Frank Lloyd Wright had been struck by the 'Raygun Gothic' of Lang's *Metropolis*. Lang's architecture was based on contemporary Modernism and Art Deco, a brand-new style in Europe at the time, and considered an emblem of the bourgeois class.

Wright's response was Broadacre City. Developed in 1935, Broadacre City was the antithesis of a city and the apotheosis of the newly born suburbia, in which each US family would be given a one-acre plot from federal land reserves.

Le Corbusier's Radiant City (1935) grew out of the Langian concept of capitalist authority and a pseudo-appreciation for workers' individual freedoms. In the Radiant City, pre-fabricated apartment houses – *les unites* – were at the centre of urban life. Les unites were available to everyone, not just the elite, and based upon the needs of each particular family.

In fiction, the dystopian image of the city endured; a place where poverty reigned, vice and crime prospered. The splendid vision of the supercivilised city, imagined in H.G. Wells' *When the Sleeper Wakes* (1899) and the artwork of Frank R. Paul, gave way to three stereotyped images of the future city.

➊ The UK has one CCTV camera for every fourteen inhabitants.
➋ Big Brother: more famous than Frankenstein?

The image of the first kind exaggerates the contrast between claustrophobic city and open wilderness. E.M. Forster's *The Machine Stops* (1909) was an early prototype, and *Æon Flux* (2005) a recent example. Here, the idea of escape from the city is dominant, one that is linked to survival itself.

The image of the second kind presents the city in decay. In this version, once proud near-utopian city-states have fallen into rack, ruin and riot, of which Samuel R. Delaney's *Dhalgren* (1975) is a legendary and enigmatic example, greatly influential on William Gibson's urban Sprawl in *Neuromancer* (1984).

The third future cityscape, exemplified in movies such as *Dark City* (1998) and *The Matrix* (1999), is the shadow city. Here an alienating and shadowy metropolis features a human experience increasingly impersonal and marginalised.

BIG BROTHER

He's as famous as Frankenstein's monster. His is the forked tongue of Newspeak. He reads your mind through his Thought Police. His realm is Room 101.

As well as providing inspiration to a generation of television programme-makers, George Orwell's masterpiece has become shorthand for almost anything thought to be oppressive and in opposition to a free and open society. Arguably, few novels have been as prophetic as *Nineteen Eighty-Four*, Orwell's haunting spectre of big government gone mad with lust for power.

Aldous Huxley in *Brave New World* (1932) had imagined the technology of titillation.

Orwell's nightmare book of 1948 had predicted the technology of control. The insidious nature of Big Brother's culture of surveillance stems from the 'telescreens' and Thought Police. In Orwell's wonderful words, 'The Beehive State is upon us, the individual will be stamped out of existence; the future is with the holiday camp, the doodlebug and the secret police'.

And in many ways we now live in an Orwellian world. Consider Newspeak. This language of political spin and euphemism is everywhere. War is 'conflict'. Civilian casualties are described as 'collateral damage'. Firing employees has become 'right-sizing'. Fixing a software problem is a 'reliability enhancement'. These are the days of Big Brother.

In *Nineteen Eighty-Four*, science's mastery of the machine is absolute. Utopia is a possibility. But inequality is preserved as a means of control. The monitoring medium of the two-way telescreen is a brilliant mutation of the idea of the all-seeing eye. But this eye is technological, not divine. In Orwell's book, Big Brother and big government replace God. Religion is replaced by science and politicised into a nightmare. As Winston Smith dutifully follows the daily exercises on the telescreen, he is at the same time observed by it.

The technology of surveillance in Orwell's fiction swiftly became fact. *Nineteen Eighty-Four* became the standard text for describing the militarisation of life. In 1954 American science historian Lewis Mumford declared the world of Big Brother to be 'already uncomfortably clear'. And US social scientist William H. Whyte cited Orwell's influence in his 1956 *The Organisation Man*, a best selling study of corporate dictatorships such as General Electric and Ford.

Orwell was troubled by the oppressive potential of science and technology. Today in the UK ('Airstrip One' in Orwell's book), there are fears that we are sleepwalking into a surveillance society. Actions of citizens are becoming increasingly monitored through the use of credit card and **MOBILE PHONE** information, and closed-circuit television (CCTV). Government monitoring of work, travel and telecommunications is also rising.

The first CCTV system was installed at Test Stand VII in Peenemünde Germany in 1942 for observing the launch of V2 **ROCKETS**. It has been estimated that there are up to 4.2 million CCTV cameras in the UK, around one for every fourteen people. In one CCTV system, pioneered in Wiltshire, England, operators can communicate directly with offenders on the spot. Orwellian indeed.

As envisaged in Steven Spielberg's *The Minority Report* (2002), within the next decade it will also be possible to scan shoppers as they enter stores. And as suggested in *Enemy of the State* (1998), we seem to have a society based both on state secrecy and a reluctance to give up the supposed right to keep information under control while, at the same time, wanting to know as much as possible.

So far, there is no Thought Police. But even that's about to change.

An American company, Oceanit, are now using sense-through-the-wall technology that can detect breathing and heart rates (and therefore supposedly possible malicious intent) from outside a building by picking up on the radio waves that humans emit. The University of Maryland's Department of

12 great androids

- Bishop – *Aliens* (1986)
- C3PO – *Star Wars Episode IV: A New Hope* (1977)
- Kryten – *Red Dwarf* (1989)
- Data – *Star Trek: The Next Generation* (1987)
- Gort – *The Day the Earth Stood Still* (1951)
- The Gunslinger – *Westworld* (1973)
- Marvin – *The Hitchhiker's Guide to the Galaxy* (1981)
- Metal Maria – *Metropolis* (1927)
- Roy Batty – *Blade Runner* (1982)
- Sonny – *I, Robot* (2004)
- T-1000 – *Terminator 2: Judgment Day* (1991)
- Terminator Series 800 – *The Terminator* (1984)

❶ *Enemy of the State* (1998).
❷ This demonstration of optical camouflage technology at Tokyo University is part space-age material, part camera trick. The coat is not see-through; a camera behind the wearer is connected to a projector in front, which displays the moving image on the coat.

Electrical & Computer Engineering is developing a technology based on the human motion analysis technologies used in health and sports science to uncover 'gait DNA', a unique 'code' of how each individual walks. This will be used alongside already well-established voice and face recognition technologies. Interestingly, the public don't seem to mind. Opinion polls, both in the US and Britain, suggest around 75 per cent of people want more, not less, surveillance.

OPTICAL CAMOUFLAGE

In the future the Peeping Tom will go high tech. No longer content with hiding up trees armed with binoculars, the twenty-first-century voyeur could well be standing next to you in the room without you even knowing. All because he is cloaked from your eyes with optical camouflage. There is nowhere to hide when you can't see your stalker. First conceived by Philip K. Dick as a 'scramble suit' in his 1974 book *A Scanner Darkly,* this high-tech hiding device has since been found all through science fiction. Arnold Schwarzenegger can't kill the *Predator* (1987) because it hides behind optical camouflage. James Bond's car is covered with it in *Die Another Day* (2002). It was the Japanese animated manga and anime *Ghost in the Shell* (1995), however, which inspired three Japanese professors to recreate the technology for real. These University of Tokyo boffins successfully created an outfit in 2003 that uses a video camera to take an image of the background and project it on the front of the suit. Although they aren't quite ready to mass-produce this chameleon-like product, it's well on its way to a shop near you.

AIR-TO-SURFACE MISSILES

It is a staple of modern warfare. We have only to turn on the news to see striking images of aeroplanes and helicopters launching missile after missile into enemy territory. The devastation caused in conflicts throughout the world is immense. Yet the concept of the air-to-surface missile owes its existence to arguably the most popular science fiction author of the 1890s. No, not H.G. Wells. Instead George Griffiths. It was this journalist who, in 1893, wrote his novel *Angel of the Revolution.* In it a

socialist and anarchist alliance uses advanced airpower to exert its dominance over the entire world. Prefiguring World War I by over two decades its descriptions of the horrors of mechanised warfare should have struck terror into the heart of mankind. Instead its description of the first use of air-to-surface missiles was to inspire military development: 'The projectiles were about two feet long and six inches in diameter, and ... there were three blades projecting from the outside, and running spirally from the point to the butt. These fitted into grooves in the inside of the cannon'. Not quite the revolution that George had in mind.

JACKING IN

Would you ever open your nervous system to a computer's virtual world? Would you ever connect yourself in a more basic autonomic way to a piece of technology so that you could exert some form of control over it? If the answer is yes then you are looking towards a future in which you can jack in. At its most extreme it is depicted in the 1999 science fiction film *The Matrix* as being accomplished by having a large metal spike directly inserted into a data connection set into the base of your skull. In this way the user is able to leave his body behind and walk as an **AVATAR** in a virtual world known as 'The Matrix'. However, the first depiction of direct organic–machine interaction in this way occurs nearly thirty years earlier.

The 1970 Robert Silverberg novel *Tower of Glass* features an artificial human called Watchman, the result of a breeding program. A supervisor on the construction of the eponymous tower, he logs in by inserting a plug into a connection in his forearm. Once in connection with the computer he directs machinery, places orders and requisitions material. His work complete, 'watchman unjacked himself'.

It was in 1981 that Vernor Vinge's novella *True Names*, in which the first **AVATAR** was described, depicted a future in which individuals 'jacked in' to a network which enabled them to see and control what their virtual bodies were doing just by using their own nervous systems. It took a further seventeen years for the first steps to be made towards making this science fiction concept a reality.

It was on 24 August 1998 that Professor Kevin Warwick of Reading University had a simple radio-frequency identification chip implanted in his arm. Once activated the chip allowed Warwick to open doors, turn on lights and control heaters by virtue of his proximity. However, Warwick could exert no conscious

⬆ An F-22 Raptor unleashes a missile.
➚ Don't try this at home: Professor Kevin Warwick with a chip implant.
➡ Trinity helps Neo to jack into *The Matrix* (1999).

control over what was happening. Still, the point had been made and this opened the way for new developments.

On 14 March 2002 Warwick had his first upgrade. A more complex chip which directly interfaced with the Professor's nervous system was inserted into his arm. With the chip not only could Warwick continue his labour-saving activities but he could also produce a signal which enabled him to directly control both an electronic arm which had been developed by a colleague and an electric wheelchair. Warwick then travelled across the Atlantic to America. Once there he logged himself in via the **INTERNET** at a lab in Columbia University. With a connection established he was able to transmit his neural signals across the world and remotely control the mechanical arm still resident in his laboratory at Reading University.

Warwick was 'jacked in'.

But the free flow of information was not all one way. Warwick and his colleagues were able to demonstrate that they could create artificial sensations in his arms by accessing the chip which had been implanted there. Using ultrasonic signals they were able to transmit information to his neural network that allowed the boffin to move blindfolded around a room without bumping into any of the hazards which littered the area.

So perhaps the answer for any clumsy person is simple. Why not get 'jacked in'; not only could you control your online activities via your neural net but you might even stop falling over the dog.

ARCADE RECRUITERS

A young lad stands in front of a large arcade game. His hands twisting desperately across the controls, he destroys enemy after enemy to move through innumerable levels. A crowd gathers around him as the game gets more and more challenging and he starts to feel the pressure. Finally, he clears the last level and is victorious. Within moments a recruiter wanders over and before he knows it the young man is whisked away to fight in a real war. Is this the plot of the 1984 science fiction film *The Last Starfighter* or a scene

from a real life arcade in the United States? In the film, young Alex Rogan is recruited to become the eponymous hero after having his fighting skills assessed by the game. Orson Scott Card's award-winning 1985 story *Ender's Game* has a similar game-playing contrivance, as does Iain Banks' 1988 *The Player of Games* and Samuel Delaney's 1976 *Trouble on Triton.*

Following such inspiration, the US Army licensed and distributed the coin-operated America's Army™ game in 2002 with the express purpose of winning recruits. Unlike many other 'shoot 'em ups', the Army's version doesn't just involve blowing people away. It is designed to reward teamwork, accuracy and use of the proper rules of engagement, and the game lacks the gore so prevalent in commercial products. Now, if only they could eliminate the gore from the real thing…

STAR WARS

On the evening of 23 March 1983 President Ronald Reagan addressed the American people in a televised statement. He argued that the world had become a dangerous place as a result of the escalating nuclear arms race with the USSR. He was therefore instituting a program to develop new technologies capable of eliminating Russian nuclear missiles in flight before they had the chance to destroy the USA – the Strategic Defense Initiative (SDI). His speech electrified the nation and sparked a massive debate.

It was quickly revealed by the US Government that there would be two types of technology associated with this Anti Ballistic Missile Shield. The first was ground-based interception devices that focused on shooting down short- and medium-range ballistic missiles by ramming them with interceptor missiles. An extension of existing technology, their inclusion in the Strategic Defense Initiative was not the really sensational part; it was the second phase of the project, which

❸ Realistic: *America's Army* computer game was launched in 2002.
❹ *The Last Starfighter* (1984).
❺ The president addresses the nation.

called for space-based weapons – particularly nuclear powered x-ray lasers – that caught the public's imagination.

Quickly dubbed the 'Star Wars' program by fans and critics alike, the vision of the weaponisation of space directly referenced the 1977 George Lucas film of the same name. With its depiction of space-based interceptors and the Death Star, an orbiting **SPACE STATION** able to fire lasers, it was almost inevitable that this nickname was going to stick. Yet the influence for this attempt to militarise space can be traced much further back. Legend has it that Archimedes created a beam by using a 'burning mirror' to reflect and concentrate the Sun's rays and incinerate the invading Roman fleet as they attacked Syracuse.

But it was the effect that science fiction had on the American psyche that has been explored by academics who have concluded that it directly contributed to public support for the SDI project. From the very first episode of the very first science fiction serial, *Flash Gordon* in 1936, a wide variety of ray-guns, death-rays and beamed weapons were deployed in space. No doubt influenced by the 1899 'heat ray' wielded by the Martians in H.G. Wells' *War of the Worlds*, these shafts of destruction would arc through the sky destroying whatever they touched.

By the time the first episode of *Star Trek* was broadcast in 1966 it seemed natural to have its starship *Enterprise* armed with such destructive weapons. These were referred to as 'phasers' and it was explained that this stood for phased-masers. This iteration of the ray-gun was directly influenced by the 1959 introduction to the public of the term and concept of the laser. The idea of mounting such devices on orbiting platforms was explored consistently in films of the sixties and seventies with notable examples including the 1971 and 1979 James Bond films *Diamonds are Forever* and *Moonraker*. The public's imagination was easily swept up in the idea of such a space-based shield – particularly when it held the hope of averting assured destruction.

As it turned out, the reality was never to live up to the rhetoric. Numerous tests were conducted in the 1980s on many types of anti-ballistic missile systems, although the public only seemed to be really interested in the x-ray laser. Extensively tested, it was never to find a military application.

On the other hand, the research accomplished as a consequence of the ground-based tests has resulted in schemes such as the current Patriot Missile System.

Still, the x-ray laser research was not a complete waste; based on the experiments conducted as part of the weapons program, new techniques in laboratory lasers for medical imaging were developed. This led to enhanced procedures for the early detection of breast cancer. So perhaps lives have been saved after all.

TRANSCRIBERS

Imagine writing a report or doing homework merely by chatting away to your computer. Having it recognise, spell and punctuate everything you say with complete accuracy. This was a vision shared by the great Isaac Asimov, who in 1953 dreamed of a future in which everybody could use an automated scribe. His future history novel *Second Foundation* features a transcriber, which translates human speech into words on paper.

Admittedly by the time Asimov was writing, computers had been produced which could replicate the human voice. It was a big leap, however, to have a machine recognise speech and take accurate notes. Imagine the time saved by such a device. Books could be written as fast as a human could speak. In fact it would take one hour and two minutes to write this book.

But whilst there have been transcription programs widely available for computers since 1995, the most popular being Dragon's *Simply Speaking*, they don't achieve the one hundred percent accuracy depicted in Asimov's work. Still it's a step closer to the science fiction vision of an **ARTIFICIAL INTELLIGENCE** which can understand and recognise enough language to carry on a spoken conversation.

AVATARS

Have you ever played a computer game as an alter ego. Perhaps you've been Lara Croft in *Tomb Raider* or the Space Marine in *Doom*. Maybe you've logged onto a message board under an alias. You could even have used an animated icon to represent yourself. If you have, you've employed an **AVATAR**. That is a fictional representation of your self used in **CYBERSPACE**.

This term for a virtual identity was first coined, then popularised, by writer Neal Stephenson in his 1992 book *Snowcrash*. 'Avatar' comes from an Indian word for incarnation and was previously used for those times when a god would take a new identity to travel upon the Earth.

As a concept it predates the **INTERNET** itself. The first true internet system went live on 1 January 1983 and the World Wide Web didn't emerge until 1989. Yet in 1981 the science fiction writer Vernor Vinge wrote the

novella *True Names*. The story featured a virtual reality where people hide their identities. The hero, Mr Slippery, fights against a mad AI, the Mailman, for control of both the virtual and real worlds. Years later, the online world has made Vinge's vision come true with 'metaverse' environments such as the phenomenally popular *Second Life* site.

HEAT-RAY WEAPONS

In *The War of the Worlds* (1898) H.G. Wells created the myth of technologically superior **ALIENS**. The thing that vividly hammers home the cosmic chain of command in Wells' book is the Martian machinery.

Their tripods tower over man. They are an unrelenting force of the void. The superior Martian machines are instruments of human oppression, and their chief weapon is the heat ray, an 'invisible, inevitable sword of heat' as Wells described it.

In early 2007 news broke that the US military had developed a revolutionary heat-ray weapon to repel enemies and disperse hostile crowds. Named the Active Denial System, the weapon projects an invisible high-energy beam that produces a sudden burning sensation in its target audience. One journalist who agreed to be blasted with the heat ray said the feeling was like a blast from a very hot oven – too painful to bear without diving for cover.

Military officials claim that, unlike its fictional Martian counterpart, the gun is harmless. It is merely a way of making enemies surrender their weapons, bridging the gap between 'shouting and shooting'. Dubbed one of the weapons of the future, it is expected that the services will add it to their tool kit as early as 2010.

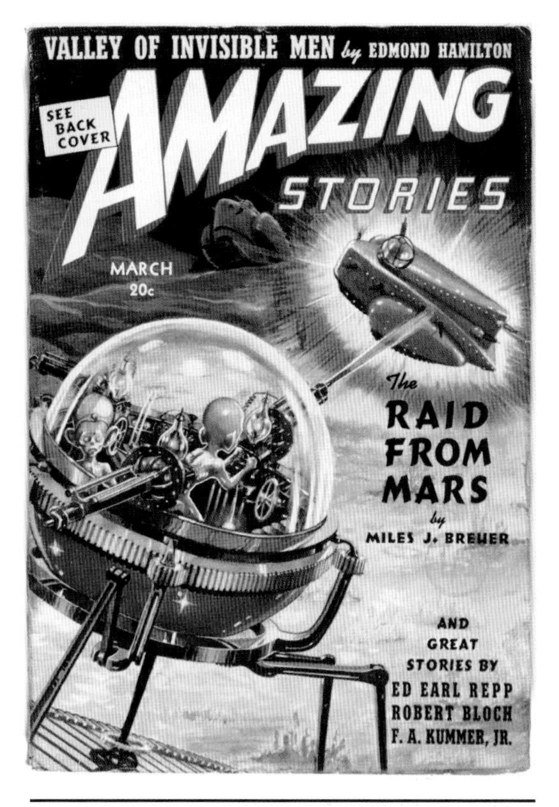

Visionary and author Isaac Asimov.

Second Life: avatars aren't bound by the laws of physics or biology. You can even be a flying spaghetti monster.

Heat-rays: handy for microwaving your foe.

SCREENSAVERS

Sitting in front of your computer working, you are suddenly called away from your desk for a moment. By the time you return the work you were busily engaged in has disappeared. Instead it's been replaced by lots of tropical fish seemingly swimming inside your screen. The screensaver has cut in after a few minutes of inactivity on the part of the computer. Originally designed to stop screen burn in cathode ray tube (CRT) displays, the technology has stayed with us even after the bulky cathode ray monitors have all but disappeared from the modern workstation. The idea that we would prefer to watch fish (or just about anything else) rather than have our work displayed in a static fashion lies with Robert Heinlein. This grand master of science fiction imagined this nifty little device, right down to the fish, in his 1961 classic *Stranger in a Strange Land*. In the novel the hero, Valentine Michael Smith, a human born and raised by Martians, returns to Earth and in the process of sharing the Martian

10 notorious science hoaxes

- Alien Autopsy film (1995)
- The Cardiff Giant (1869)
- Clonaid and baby Eve (2002)
- Hwang Woo-Suk affair (2004–06)
- In His Image: The Cloning of a Man by David Rorvik (1978)
- N-rays (1903)
- Piltdown Man (1908–15)
- Charles Redheffer's Perpetual Motion Machine (1812)
- The Alan Sokal/Social Text affair (1996)
- The Turk Chess-playing Machine (1770–1854)

philosophy revolutionises Earth culture. The screensaver hasn't quite achieved the same level of influence but it does provide us with pretty pictures of blue neon tetras and black guppies. Isn't that enough?

Personal video players

Sitting on a train bored with your journey, you could pull out a book or read a paper. But then you remember that you downloaded that programme you missed on television last night. Pulling out your personal video player, you use your thumb to scroll past the gigabytes of music, the folders full of photos and the electronic books until you get to videos. Selecting the episode you want, you settle back into your seat to enjoy the journey.

H.G. Wells' 1899 story *When the Sleeper Awakes* uncannily predicts this technology. Owing a debt of inspiration to Washington Irving's *Rip Van Winkle*, a dystopian tale of a man who wakes from a two-hundred-and-three-year coma to find that the interest on his assets has made him the world's richest person, it features something which would seem strangely familiar in the hands of any of today's youth. Opening a small flat box he sees that inside 'on the flat surface was now a little picture, very vividly coloured, and in this picture were figures that moved. Not only did they move, but they were conversing in clear, small voices. It was exactly like reality.'

The internet

Nobody owns the internet. Sure they can purchase their own little part of it, but nobody owns it all. This global collection of networks, large and small, is just too immense for any group or individual to possess. Every time you log on you are skipping your way between different networks connected together in many different ways. It is this loose affiliation of connections that we call the internet – a term which comes from the idea of interconnected networks.

Long before we had the technology to make this a reality, science fiction was exploring the concept behind what we recognise as the internet today. It was in 1937 that perhaps the most influential science fiction author of all time, H.G. Wells, released his work *The World Brain*. In this essay Wells explored the idea of an ever evolving, rapidly updating compendium of knowledge spread across the face of the planet: 'There is no practical obstacle whatever now to the creation of an efficient index to all human knowledge, ideas and achievements, to the creation, that is, of a complete planetary memory for all mankind. And not simply an index; the direct reproduction of the thing itself can be summoned to any properly prepared spot.' Sound familiar?

The difference is of course that for Wells the means by which the knowledge was accessed was via microfilm.

Wells was developing ideas which had already been pioneered by one of the founders of information science, Paul Otlet. This Belgian academic and peace activist talked about concepts akin to hyperlinks, search engines, remote access and social networks although, as with Wells, his means of accessing the information was based around not electronic technology but the rather more mundane idea of paper documents. His work was hugely influential between the two world wars.

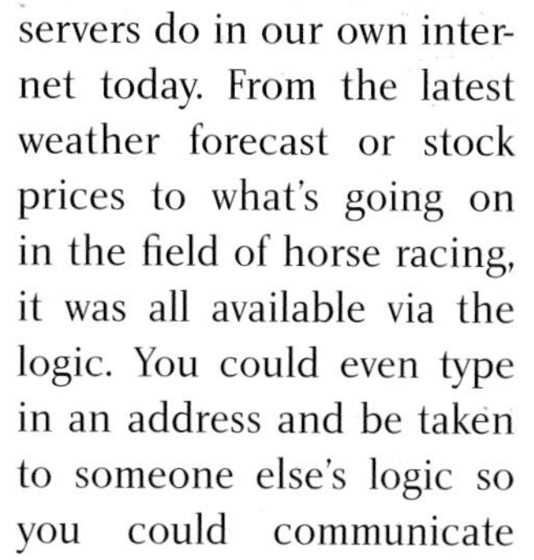

It was not until just after the Second World War, by which time Otlet had died, that a vision of a technological network similar to the internet was first depicted. In 1946 the science fiction author Will F. Jenkins, who wrote under the pseudonym Murray Leinster, had his story *A Logic Named Joe* published. Leinster's depiction involved individuals sitting in front of 'logics'. These were large television screens with keyboards attached. Using an innovation referred to as a 'carson circuit' these logics were able to punch up different sites across the logic network via different sets of 'tanks' which acted as servers do in our own internet today. From the latest weather forecast or stock prices to what's going on in the field of horse racing, it was all available via the logic. You could even type in an address and be taken to someone else's logic so you could communicate with them in the same way that we use email and instant messaging systems today. The carson circuit in each logic was its address, and although it was a physical item it acted in the same way that a URL (Uniform Resource Locator) or web address does now.

The 1950s saw developments in computer networks, although these predominantly focused on the idea of allowing computers on a single network, connected by long data

The Apple video iPod.
The internet relies on hundreds of thousands of miles of cabling and wiring.
In *The World Brain* (1937) H.G. Wells explored the idea of a constantly evolving worldwide compendium of knowledge and information.

lines, to communicate via a mainframe. It was not until 1962 that the newly appointed director of the US Government's ARPA (Advanced Research Projects Agency), J.C.R. Licklider, asked that three terminals all connected to different networks but housed at ARPA be made to talk to each other that the first faltering steps down the road to the internet were made. The first message sent across the ARPANET took place in January 1969. It was to be the word 'Login'. However, the system crashed after the first two letters. The first internet message was therefore 'Lo'.

TELEFACTORS

Have you ever used one of those kids' toys which are effectively a mechanical hand remotely operated from the end of a stick? Perhaps you have used a litter picker? If you have, then you have operated a telefactor. You have used a basic remote manipulator to accomplish a task which could otherwise be done by human hands.

More advanced models are commonly used throughout industry, particularly in hazardous environments like nuclear power stations. It enables the operator to move dangerous materials in safety. These devices are on the frontline in many difficult and perilous environments. Having many applications, the remote arm on the space shuttle can be considered a famous (and rather large) example. Commonly called Waldos, they are so named from the 1942 story *Waldo* by Robert Heinlein in which the eponymous hero, who suffers from a muscle-wasting disease, uses a telefactor in his orbital laboratory. Since its conception, it has been seen in numerous science fiction films. A basic

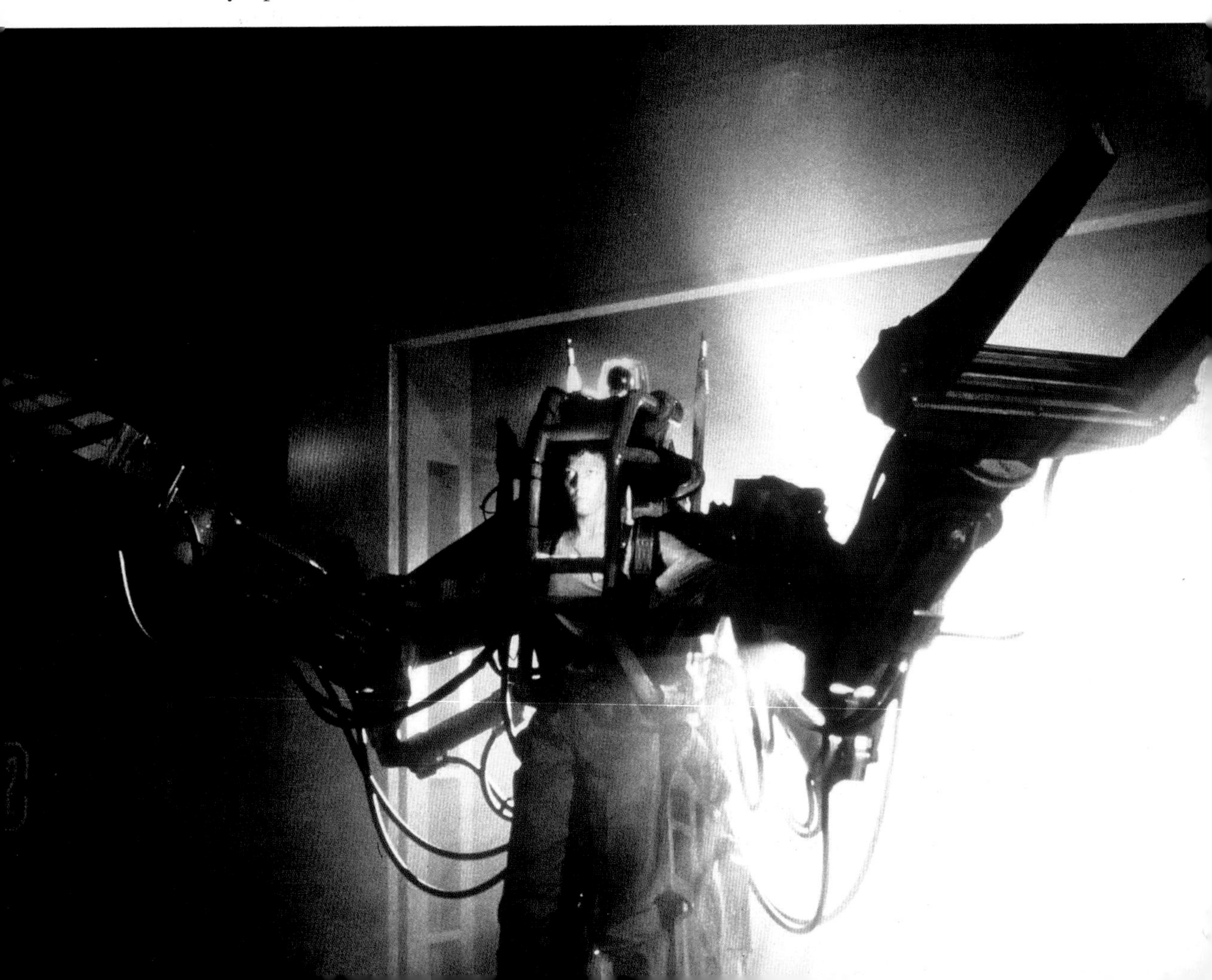

laboratory model appeared in *The Andromeda Strain* (1961), whilst in *Aliens* (1986) the heroine, Ripley, dons a complete exoskeleton power-loader Waldo to fight off the alien queen. Just what you need to clear out the spare room.

PERSONAL WIND POWER

At a time when we are all concerned with saving the planet and reducing our carbon footprint, more and more innovative ways are being found to reduce our reliance on fossil fuels. One method which is becoming increasingly popular is mounting a micro wind turbine on top of your house. This mini-windmill generates enough electricity to run your energy-efficient light bulbs. Although wind power has been used for millennia, it was John Jacob Astor IV's 1894 *Journey in Other Worlds* which envisaged a future in which we are all responsible for our own electricity generation. Following the 1887 success of wind power pioneers, Professor James Blyth of Anderson's College, Glasgow, and the American Charles F. Brush in Cleveland, Ohio, Astor speculated that a day would arise when the huge frames of his time (Brush's rotor was seventeen metres in diameter) would be replaced by more compact personal models. Thus each would have 'a windmill on his roof for light and heat; then, the harder the wintry blasts may blow the brighter and warmer becomes the house, the current passing through a storage battery to make it more steady'.

ROCKETS

Firstly, there was the firework. Many moons ago, and long before the days of the Darwin Awards, there lived a man named Wan-Hoo. Wan-Hoo was a minor Chinese official of the Ming Dynasty. He was also the world's first astronaut. Allegedly.

Legend has it that, early in the 1500s, Wan figured he could launch himself into outer space. Cunningly using China's advanced firework technology to his advantage, Wan built his spaceship. A chair. To this chair Wan fastened 47 large rockets. Using what influence he could muster within the Dynasty, Wan called up 47 assistants. Each willing assistant, armed with a torch, was charged with the task of rushing forward and lighting one of the long fuses.

On the day of lift-off, the finely attired Wan climbed onto his rocket chair and his 47 aides lit the fuses. The assistants hastily ran for cover. There was a tremendous roar, and a huge explosion. The smoke cleared. The rocket chair was gone. Wan was never seen again.

Now this tale was first reported, not in ancient Chinese manuscripts, but in *Rockets and Jets*, and written by an American author, Herbert S. Zim, in 1945. The account was later introduced into China via translation. Nonetheless, the legend lives on. The International Astronomical Union named a

◀ That's a Big Waldo, Ellen: from *Aliens* (1986).
▲ Small-scale wind turbines rarely generate as much power as their owners would like.

➔ Fly me to the Moon: Georges Méliès' *Le Voyage Dans la Lune* (1902).
➔ A rocket from Jules Verne's *From the Earth to the Moon* (1865).
➔➔ The Korean border, where the SGR-A1 sentry gun is deployed.

crater on the far side of the Moon after Wan-Hoo when the far side was first photographed in the 1960s.

For centuries after Wan-Hoo, writers and engineers grappled with the idea of propulsion. In his 1634 voyage to the Moon, *Somnium*, Johannes Kepler had his hero spirited away by demons. Francis Godwin, in *The Man in the Moone* (1638), shipped his protagonist Moonwards using birds he called 'gansas'. And given that the Sun seems to 'draw up' dew-drops, Cyrano de Bergerac, in his *The States and Empires of the Moon* (1657), suggested that one might fly by trapping dew in bottles, strapping the bottles to oneself, and standing in sunlight. Pure genius.

Jules Verne sent the circus into space. In his 1865 book *From the Earth to the Moon* (*De la Terre à la Lune*), Verne's preferred method of propulsion was the cannon. Or, rather, the Columbiad: a large calibre, smoothbore, muzzle-loading cannon. It was capable of firing heavy projectiles at both high and low trajectories. Verne picked a very high trajectory, and an ambitious target – the Moon.

His chosen cargo, three affluent members of a post-American Civil War gun club, is launched in a projectile-cum-spaceship from an enormous sky-facing Columbiad. Verne had included some surprisingly accurate calculations on the requirements of the cannon, though a far longer muzzle would have been needed for the cargo to reach 'escape velocity' (the speed required to break free from Earth's gravitational pull). His Moon-landing scenario also proved a little bumpy.

Verne's tale bears striking similarities to the Apollo program. The Apollo 11 command module, with a crew of three, was called *Columbia*. The dimensions of Apollo command service modules are very close to that of Verne's projectile, and their chosen launch site is also Florida. Verne had realised, as NASA did later, that a launch is easier from near the Earth's equator.

But in 1881 Nikolai Kibalchich finally invented the principle of the modern **ROCKET**. As he waited in prison to be executed for the assassination of Alexander II, Nikolai, a Russian revolutionary and explosives expert, dreamt up a new method of propulsion: gasses produced by slow burning explosives escaping through a nozzle.

Nikolai became an inspiration for Russian rocket pioneer and science fiction writer Konstantin Tsiolkovsky. The road to **STAR WARS** had begun . . .

SENTRY GUNS

In a classic scene from the movie *Aliens* (1986), the aggressive xenomorphs are barrelling down a corridor at a group of marines. As they turn a corner they are confronted by a robotic sentry gun. This tripod-mounted remote device immediately tracks the individual **ALIENS** before eliminating them in a hail of automatic gunfire. Originally conceived by the fertile mind of Michael Crichton for his 'killer germs from outer space' novel, *The Andromeda Strain* (1961), the true home of

the sentry gun has come to be the many first person shooter video games in which it features. Yet worryingly, recent developments have also resulted in similar devices being deployed in large numbers in one of the most dangerous places on Earth. Not content with millions of mines, miles of barbed wire and thousands of men, the border between North and South Korea has now gone high-tech. The new South Korean Weapons Grade Robot can 'detect, raise the alarm, and provide suppressive fire', said Lee Jae-Hoon, deputy minister of commerce, industry and energy in an interview in 2006. Equipped with visual and infra-red detection capabilities, the Samsung-developed SGR-A1 sentry gun can spot moving objects over two miles away during the day and half that distance at night. Via pattern recognition, it can supposedly distinguish between humans, cars and trees, although we don't recommend you put it to the test.

THE ATOMIC BOMB

'...And these atomic bombs which science burst upon the world that night were strange, even to the men who used them.'

So quoted H.G. Wells in his prophetic 1914 novel *The World Set Free*. Wells' book was the first to christen the 'atomic bomb', and his story led non-stop to Hiroshima.

Fat Man: a replica of the atomic bomb that was dropped on Nagasaki on 9 August 1945, three days after Hiroshima.

Leo Szilárd, a leading light of the Manhattan Project.

The 18,000 metre-high mushroom cloud that followed the detonation of the atomic bomb over Nagasaki.

By the dawning of the twentieth century it was clear that some form of atomic energy was responsible for powering the stars. Scientists such as the great nuclear physicist Ernest Rutherford and his co-worker, Frederick Soddy, realised that atoms were the seats of enormous energies. And even though Rutherford is alleged to have suggested 'some fool in a laboratory might blow up the universe unawares', both Rutherford and Soddy trusted nature to 'guard her secret'. H.G. Wells begged to differ.

His timetable for the development of nuclear capability is unnervingly far-sighted. In *The World Set Free*, the 1950s scientist who uncovers atomic energy realises there is no going back. Nonetheless, he feels, 'like an imbecile who has presented a box of loaded revolvers to a crèche'. Wells also envisaged a world war that would take place in 1956, with an alliance of France, England and America against Germany and Austria.

Wells' book predicts a holocaust, in which the world's major cities are annihilated by small atomic bombs dropped by aeroplanes. This is no mere guesswork. Wells' weapons are truly nuclear; Einstein's equivalence of matter converted into fiery and explosive energy triggered by a chain reaction.

Wells' visionary novel was the guiding light for brilliant Hungarian physicist Leo Szilárd. After reading *The World Set Free* in 1932, Szilárd became the first scientist to seriously examine the nuclear physics behind the fiction.

On reading an article in *The Times* by Rutherford rejecting the idea of using atomic energy for practical purposes, Szilárd was incensed. His fury, fused with his legendary quick wit, enabled Szilárd to dream up the very idea of the nuclear chain reaction that Rutherford denied while waiting for traffic lights to change on Southampton Row in Bloomsbury, London. One year later Szilárd filed for a patent on the concept.

Szilárd became the driving force behind the Manhattan Project. It was his idea to send a letter in August 1939 to Franklin D. Roosevelt outlining the possibility of nuclear weapons.

Wells' fiction became factual terror over Japan. As the 320,000 inhabitants of Hiroshima were waking up, the bomb burst over the city. Thousands were slain in a second; they perished in a 'heat death', vapourised by light and energy. Ghostly shadows on nearby walls their only remains, they were the lucky ones. Victims further from the blast were blinded, or had their skin and hair ablaze. Later they would lose the white blood cells needed to fight the escalating disease.

Szilárd had hoped that President Truman would merely 'demonstrate' the bomb, not use it against cities as in Wells' *The World Set Free*. But as the war raged on, scientists lost the power over their research.

The Manhattan Project's lead scientist, Robert Oppenheimer, mulled over the 'atomic bomb' first imagined by Wells. Oppenheimer spoke for many physicists when he said, 'In some sort of crude sense which no vulgarity, no humour, no overstatement can quite extinguish, the physicists have known sin; and this is a knowledge which they cannot lose.'

MONSTER

We like being frightened. That's why we go on roller coasters and that's why we enjoy watching monsters. The 'fight or flight' response, evolved in our distant ancestors, is hardwired into our sympathetic nervous system. Thus when we are frightened our brains release the hormone norepinephrine, which raises our heart rate and makes us more alert.

Science fiction didn't invent monsters. They have always been part of the stories that we tell. The ancients Greeks had many fine examples, as did the Indian sub-continent. In fact science fiction has often turned to science to provide a gloss of explanation for the 'classic' monsters. Weird wolf-men become the victims of a disease known as lycanthropy. Zombies become the result of viruses escaped from the laboratory. Mother Nature herself becomes both a monster in the form of severe weather and a source of monsters. Thus we have giant apes scaling skyscrapers, ambulatory plants seeking to blind us and flocks of birds intent on revenge.

Yet it seems that science fiction reserves its greatest treatments of the monster for the definitive terror: ourselves. The monster that looks back out at us when we gaze into the mirror. It seems that our reputation as the predator par excellence is perhaps deserved after all. For what we do to ourselves is more horrifying than anything that can be done to us. The monster story becomes an allegory of our own fears and concerns over the world that we have created. Thus the rampaging giant Godzilla dominating downtown Tokyo is depicted as the result of radioactive mutation. Our own fear of fission wrought large in lizard-like form. Yet even then it stands on two legs waving two arms.

There is nothing that is truly alien in science fiction. For to be alien is to be incomprehensible and how could we relate to that? So we ascribe to our monsters, if not human forms, then human desires and motivations. The monster, whether it be alien or killer android, is always a grotesque and warped version of our own debased intentions. We have dressed up those corrupt desires in the most elaborate of disguises. We have concealed them beneath layer upon layer of civility and custom until they became known by their present form – civilisation.

But science fiction reminds us that lurking beneath the mask waits the creature ready to erupt. Its many books and films remind us that the age of progress and civilisation that we have created is a fragile masquerade. Like a house of cards, our cunning construction could fall apart at a moment's notice. So science fiction deploys armies of bug-eyed monsters from space, primeval monsters from below and ancient horrors from within to tear down this precarious edifice.

Yet for every threat there is a response, and ours tend to come clad in spandex. Rippling muscles at the ready, the world of science fiction is protected by the superhero. Armed with laser vision, the power of flight, super-strength and gallons of righteous indignation, these pneumatic Narcissuses come to our aid. Whether benevolent mutants, aliens or accidental heroes, these monsters come not from the id but the ego. They stand as totems to our own ability to overcome obstacles and negotiate peril. Always at hand to deliver the obligatory lecture on how naughtiness never triumphs the heroes most often conspire to allow the monster to defeat itself.

In the same manner, we too are most often the architects of our own destruction. Science fiction reminds us of the consequences of our actions. For whilst it may seem like a good idea to test out that new super-serum on yourself, Doctor X, you've only got yourself to blame when you turn into a flesh-devouring, maiden-despoiling fiend. To put it bluntly – just because we can, doesn't mean that we should.

So, within these entries you will find some cautionary tales of science gone wrong and learn of some real-life developments that have worrying potential. You will find monsters that tower over us, making us realise just how insignificant we are. Yet just as unnerving is the minuscule mayhem caused by legions of microscopic and nano-scale beasties intent on turning our insides out. You will discover how some of the once seemingly far-fetched things that we do to ourselves today were anticipated. And still the relentless march of the monster continues. Whether it is lurking in some innocuous form ready to transform from your car into a towering oversize ogre or carried in the viral-laden touch of a neighbour, these threats are here to stay.

MUTANTS

The science fictional obsession with mutants has long antecedents. From the subterranean morlocks of Wells' *War of the Worlds*, via the gargantuan *Godzilla*, right up to the present day *X-Men*, it is deviation from the norm that attracts us. Yet mutation is in itself an established biological process that lies at the heart of science. Without mutation none of the plants and animals on Earth would exist in their present form. It is easy to dismiss mutant animals as monsters. Witness *Mant* or countless numbers of mutant snakes and spiders in innumerable horror B-movies like *Python* (2000).

It is the dark side of mutation, the deliberate mutation of creatures or, worse still, the mutation of ourselves into something different that seems to grip our attention. The history of science fiction is littered with casual monsters wrought through the machinations of atomic power or man's deliberate experimentation. Yet science seems intent on following in fiction's footsteps. Until recently its study of mutations was largely confined to the humble fruitfly. However, more recently a new subject has been identified. The new animal of choice is the zebrafish, *Danio rerio*. Using radiation, chemicals or viruses, scientists intentionally generate mutations in these colourful little creatures in order to discover their genetic repertoire. One experiment even resulted in a fish that exploded when exposed to light. All fiction has done is extrapolate a future in which we are the next test subject.

History provides us with some examples of difference within the human form. From conjoined twins to those affected by cyclopia (one eyed individuals named for Homer's famous monster), ectrodactyly (lobster-claw syndrome) or sirenomelia (the so called 'mermaid's syndrome', where the lower limbs are fused together), numerous examples litter the history books. One of Darwin's most insightful discoveries about heredity was the result of pondering a family whose bodies were entirely covered in hair (congenital hypertrichosis lanuginosa) and had been kept as amusements at the Burmese royal court for four generations.

An incredibly popular mutation depicted in fiction is that of the 'strong mutant'. From the *Incredible Hulk* (formed via gamma radiation) to *Juggernaut* from the *X-Men* series (a result of genetic mutation), the stronger than normal human is one of the most enduring constructs. In 2004 science caught up when the *New England Journal of Medicine* reported on a German boy who had been born with a genetic mutation that blocked the production

⊖⊖ Spencer Tracy starred in Hollywood's take on Robert Louis Stevenson's classic story of the monster within, *The Strange Case of Doctor Jekyll and Mr Hyde*.
⊖ Somebody made him angry – The Incredible Hulk.
⊖ The conveniently adaptable zebrafish.
⊖ Tin Man blows his top in *The Wizard of Oz* (1939).

of a protein called *myostatin* that limits muscle growth. At the age of four, the boy proved capable of holding his arms outstretched whilst grasping seven-pound weights in each hand, something most of us aren't able to do as adults. This terrific toddler may well grow into a powerful figure, the likes of which we have only seen thus far in the movies. Yet for all of this he is not a monster. Just someone slightly different from the norm. In the 2007 *Masters of Science Fiction* episode 'The Discarded', Brian Dennehy's mutant character explains that it was not a disease, as authorities had claimed, that caused Earth to reject its mutants, but the fact that they looked different.

Science classifies mutants in the animal and plant kingdoms by defining one form as the 'natural' strain and all others as aberrations. We have been reluctant to do so for ourselves, so that instead we have constructed the conceit of one homogenous human race. It is our diversity that makes us who we are. We are all mutants. Science fiction has been working for a long time to prepare us to accept the fact.

CYBORGS

Walking down the street you would never know that the people you nod to are cyborgs. Yet increasing numbers of us are becoming human-machine fusions. Glasses, contact lenses and hearing aids do not make us cyborgs. These are just tools we use in order to improve the way we function. However, pacemakers, artificial joints, insulin pumps and cochlear implants can all move us into the realm of science fiction. The term itself dates from an article by theorists Manfred E. Clynes and Nathan S. Kline in the September 1960 edition of the academic journal *Astronautics*. In their article 'Cyborgs and Space' they coin the term 'cyborg' (combining cybernetics with organism) to describe man-machine combinations expected to be used in the space program. However, the concept had been explored long before that.

An unlikely progenitor is the Tin Man from Frank L. Baum's *Oz* series. This hero and companion to Dorothy on the Yellow Brick Road was conceived by Baum for the 1900 classic tale *The Wizard of Oz* as a lumberjack named Nick Chopper. Betrothed in marriage to a munchkin girl named Nimmie Amee, the

Wicked Witch of the East conjures a magic axe that chops off his limbs one by one. He gradually gets these replaced with tin versions, becoming a cyborg en-route to his final tin form.

A less well-known example was Nyctalope, the cyborg **SUPERHERO** of the French science fiction series which began with the 1908 novel *L'Homme Qui Peut Vivre Dans L'eau* (The Man Who Can Live in Water). Nyctalope has artificial eyes and an artificial heart which he uses to good effect to fight a **MUTANT** shark-man in the book. Mostly human, Nyctalope stands in comparison with Deidre the dancer heroine of C.L. Moore's 1944 short story *No Woman Born.* Almost burnt to death, Deidre's brain is inserted into a supple artificial body so that she may continue to dance – perhaps the most extreme form of cyborgisation.

It was the 1972 novel *Cyborg* by Martin Caidan that was to have a lasting impression on the public psyche, however. It introduced us to a test pilot, Steve Austin, who after a near fatal air crash has large parts of his body replaced with bionic limbs. Later adapted into the 1974–1978 popular television series *The Six Million Dollar Man,* the book contrasts with the series' emphasis on spying by focusing on the recuperation and rehabilitation of the hero.

In 2000 Dr Miguel Nicolelis, a neurobiologist at the American Duke University Medical Centre, taught a monkey to manipulate a robotic arm using thoughts transmitted through electrodes implanted in its brain. This opened up the possibility of such advanced prostheses that they could be called bionic.

On 21 May 2001, Jesse Sullivan of Dayton, Tennessee, USA, came into contact with a live cable on his job for a power company. Both of his arms had to be amputated at the shoulder because his burns were so severe. He faced a future in a body brace to which two artificial arms were connected controlled by chin levers. However, a team from The Rehabilitation Institute of Chicago were able not only to fit Sullivan with innovative bionic arms but also

⊙ With his two bionic arms, Jesse Sullivan is the world's first cyborg.
⊙ Pin up of the 1990s … Dolly the Sheep.
⊙ Lynda Carter as Amazonian Wonder Woman.

to perform the necessary nerve and muscle grafts which now enable him to control his prostheses. In 2002 Jesse Sullivan became the world's first bionic man. Truly the age of the cyborg has arrived.

ANIMAL CLONING

Aside from Philip K. Dick's 1968 novel *Do Androids Dream of Electric Sheep*, the love affair between sheep and science and science fiction had admittedly been modest. However, all that changed on 5 July 1996 when the world's first artificially cloned mammal was born. Dolly, as she was known to her friends, was a female *Ovis Aries* Finn Dorset sheep. She was made from DNA taken from a grown-up female sheep's teat, and was named after the prodigiously mammiferous country and western singer Dolly Parton.

Apart from findings that may help research into the cloning of human tissue, another reason to clone animals is to preserve endangered species. In 2000 a gaur (an endangered sub-species of wild ox) was successfully cloned. The next logical stage would be to construct a park of endangered and or extinct species. Sound familiar? Michael Crichton conjures up just such a scenario in his 1990 novel *Jurassic Park* which featured cloned dinosaurs running amok. Scientists at the Mammoth Creation Project run by Kinki University in Japan hope to establish a Pleistocene Park with Woolly mammoths roaming free. Not as frightening as T-rex but slightly more frightening than sheep – although the 2006 comedy-horror film *Black Sheep* did feature well-meaning environmentalists accidentally infecting a flock of sheep with a genetically modified lamb foetus, resulting in murderous Dollys eating the inhabitants of New Zealand.

SUPERHEROES

What does the future hold for Man? What will Man one day become? Such questions have busied the fertile brains of science fiction writers, filmmakers and artists ever since Darwin.

German thinker Friedrich Nietzsche had floated the notion of the '*Übermensch*' ('superman', 'over-man' or 'super-human') in his 1883 book *Thus Spoke Zarathustra*. Nietzsche's idea of the *Übermensch* was of a being seeking to move 'over' or beyond its state of being to a greater 'stature'.

No other symbol in science fiction has evolved as dramatically as the 'super-man'. From the most infantile form of human wish-fulfilment to the more sophisticated anti-hero, the superhero has become an ingenious metaphor of our aspirations and fears for future science.

Twenty-first century cinema is replete with this 'over-man'. The genre of superhero fiction began in 1938 when pulp fiction writer Jerry Siegel and artist Joe Shuster unveiled *Superman*. Since then, superheroes have broken into radio, television and books, and they increasingly seem to be the staple of the modern, CGI-dominated blockbuster movie.

Unlike heroes of the past, such as *Tarzan* or *Zorro*, the modern superhero is a different breed. Sometimes they are highly skilled with easy access to superscientific gadgetry, such as *Batman*, dreamt up in 1939 and with his own comic from 1940. Other superheroes possessed inhuman powers, derived from some chance interaction with a scientific world.

Superman, of course, is an **ALIEN**. His power is sourced from the mere fact that he was born on the alien planet of Krypton. A bite from an irradiated arachnid spawns changes in Peter Parker's body, giving *Spiderman* his superpowers. And *The Fantastic Four*, the first superhero team created by writer Stan Lee and artist Jack Kirby for *Marvel Comics* in 1961, gained their superpowers after exposure to cosmic rays during a space mission. Of the four, *Mr Fantastic* is a science boffin, capable of stretching his shapeshifting body; the *Invisible Woman* can make herself invisible, of course, and project powerful force fields; the *Human Torch* can throw flames and fly; and the monster-like *Thing* possesses superhuman stamina and strength.

In the same way that Darwin led science fiction to the alien, theories of evolution have given writers an imaginative framework for stories of superhumans. But there is a difference, of course, between Darwin-induced narratives of 'fitter' humans, and superheroes whose

⬆ Superman: prissy and sexless?
➡ Lookout! Here comes the most popular superhero of them all.
⬅ Charles Darwin's cousin, Francis Galton.
⬅⬆ Gulliver amongst the Houyhnhnms.

creative evolution is often instantaneous, and whose new-found powers may well be passed on to their offspring. If they ever had sex.

Creators of supermen had originally been surprisingly shy to make their heroes outright villains. Critical of the contemporary human condition, it seems many writers have opted for 'progress', crediting themselves with a proto-superhuman perspective. It is very tempting to love the notion of the superhero if we believe we may become superhuman ourselves.

Sadly, all this virtuous 'progress' made some superheroes rather dull. *Superman* was prissy and sexless. *Captain America* was unable to become intoxicated by alcohol. So artist Jack Kirby became the presiding genius of a new anti-hero format for superheroes in the 1960s. His creations had sex. They had neuroses. They behaved badly. Sometimes, they even chose to become supervillains instead.

So developed the more sophisticated superhero of the graphic novel. In landmark publications such as Frank Miller's *Batman: The Dark Knight Returns* (1986) and Alan Moore's *Watchmen* (1986–7), a new creative force was born. These novels confront the question of what human society might be like if science or pure chance granted us superhero status. How complex, corrupting and weary it all may prove.

Although superheroism would still be worth a go, don't you think?

eugenics

How do we improve ourselves? Perhaps it's a case of taking extra exercise, eating healthy food or reading a book. People throughout history have advocated a social philosophy that says that the answer lies in intervening in society by improving the human race as a whole through bloodlines. Following the principle of heredity (that we

inherit certain things from our ancestors) it should be possible to raise the general profile of the entire human race. Plato in his 360 BC seminal text *The Republic* was one of the first to advocate measures (including abortion and infanticide) to achieve this improvement. Opposition was led by Hippocrates (the founder of medicine) and incorporated into the Hippocratic Oath that doctors still take.

It was a work of proto-science fiction to first explore what a society based on eugenics would be like. Jonathan Swift in his 1726 classic *Gulliver's Travels* finishes his work with his hero arriving in the country of the Houyhnhnms. These creatures, which are identical to horses, operate a eugenics program involving the selective breeding of their human slaves referred to as 'yahoos'. Initially Gulliver himself is mistaken for one of these 'degenerates' and it is only by virtue of his intellect that he is able to convince the equine masters otherwise. Swift's satire on this approach was also to find voice in his 1729 pamphlet *A Modest Proposal* in which he satirised that the logical conclusion of such policies would be for the starving poor to eat their own children.

The theories invoked by Swift in *Gulliver's Travels*, when discussed, did not have a name.

That was left to Charles Darwin's cousin Francis Galton. Impressed by Darwin's work on evolution, Galton extrapolated the approach into human heredity and coined the term 'eugenics' from the Greek words *eu* which means good and *gen* which refers to birth or race. Galton's philosophy is based on a theory of inequality – that some are born better than others. This philosophy would frequently be used to justify opposition to miscegenation (mixing races). H.G. Wells has been accused of holding similar views and, whilst there is evidence that he advocated the construction of a scientific elite which was eugenic in its approach, Wells was a prominent advocate against racism and portrayed a future in his 1895 work *The Time Machine* in which racial segregation (into the Eloi and Morlocks) would lead ultimately to human extinction. Instead he argued that competitive selection would eventually raise all mankind to a higher level of comprehension.

The most chilling approach to eugenics was embodied by the Nazi regime during the 1930s and 40s. Hundreds of thousands of men, women and children were forcibly sterilised so as to prevent them from passing on their genes. Prefiguring this and the horror of the Holocaust was a 1937 science fictional dystopia called *Swastika Night* by the British novelist Katherine Burdekin. In this future history, the Nazis came to dominate the world, imprisoning women and forcing them to become little more than breeding machines for their own Aryan super race. Seemingly prescient of the horrors that were to be revealed after World War II, Burdekin's only major miscalculation was the portrayal of the eugenics of the future as being achieved with GENETIC ENGINEERING, rather than with breeding programs and genocide.

Eugenics on the march: soldiers of the Nazi regime.

'Look into my eyes' – Kenny Craig is spellbinding.

Author of *Engines in Creation*, Eric Drexler.

MIND CONTROL

The science-induced fictional obsession with controlling the minds of others proliferated in the US early in the nineteenth century, following the discoveries of Franz Anton Mesmer. He developed what he called 'animal magnetism' and the evolution of his ideas and practices led to the development in 1842 of hypnosis, hence the term to 'mesmerise'.

Many science fiction tales have featured mind control, both through natural and artificial means. Ford McCormick's *March Hare Mission* (1951) imagined a mind control drug, 'nepenthal', which wiped clean the recipient's short-term memory.

Arthur C. Clarke described a mechanical method for manipulating the mind in *Patent Pending*, a 1954 story that envisaged the recording of memory and thoughts for later use. Michael Crichton's *The Terminal Man* (1972) experimented with a similar idea. The novel's neuroscientists attempt brain control through electrode implants.

In real life, methods of coercive persuasion (brainwashing) have been used to reverse a person's convictions. Through the use of different agents of manipulation, natural and mechanical, control can be achieved of all that a subject experiences. They can even penetrate what we might call inner conscience. It was through such brainwashing that 'messiah' Jim Jones led 913 of his followers to commit mass suicide at his commune in Guyana in 1978.

NANOMONSTERS

The world could be destroyed in many ways. Blown up with **ATOMIC BOMBS** or eaten by **BLACK HOLES**. However, one of the more extreme methods would be for it to be consumed by nanomonsters. This was brought to the attention of the science community by molecular nanotechnology pioneer Eric Drexler in his 1986 book *Engines of Creation*. Drexler outlines a scenario in which rapidly reproducing nanomachines voraciously consume all matter on the planet, reducing it to a lifeless shell. He engagingly dubbed these tiny terrors 'grey goo'. The 'goo' or 'exponential ecophagic nanogrowth' scenario has since inspired numerous science fiction stories of nanotechnology running amok.

Although in most cases it is defeated by the heroes at the last moment, in Will McCarthy's 1999 novel *Bloom* not only does the grey goo win and overwhelm the inner solar system, but it is also well on its way with the rest of the planets.

It was, however, Stanislaw Lem in his 1959 short story *Ciemnosc i Plesn* (Darkness and Mildew) who first speculated about such a disaster. In this tale samples of an artificial lifeform which uses nuclear energy breaks out of confinement. The light sensitive life is catalysed by mildew and once it finds a nice dark space it starts to grow exponentially. So check under the sink – that damp cupboard could contain the end of the world.

MULTIPLE PERSONALITIES

It was the 1886 science fiction novella *The Strange Case of Dr Jekyll and Mr Hyde* by Robert Louis Stevenson that started the public, and science fiction's, obsession with multiple personalities. In this seminal work Dr Henry Jekyll finds a way of pharmacologically separating his good and evil natures. He is left with two distinct personalities: the 'waking' version of Henry Jekyll and the murderous Hyde. With many later incarnations and versions, the notion of a 'Jekyll and Hyde' character has found its way into common parlance.

Other notable 'multiples' include Rose and Thorn, the two halves of a botanist adversary of Batman's who first appeared in *Flash Comic* in November 1947, and the protagonist Bob/Fred in Philip K. Dick's 1977 *A Scanner Darkly* (and its 2006 film remake).

However, it was the influence of a real-life account of multiple personalities in the 1973 book *Sybil* (later made into a 1976 film starring Sally Field) that was to really influence science. Following widespread public interest, in 1980

↑ *Dr Jekyll and Mr Hyde* – a study of human duality.

→ Playing God? Science fiction has reflected public anxiety over genetic engineering.

12 mad doctors

- Dr Benway – *The Naked Lunch* by William S. Burroughs (1959)
- Dr Emmett 'Doc' Brown – *Back to the Future* (1985)
- Dr Caligari – *Das Kabinett des Dr. Caligari* (1920)
- Dr Victor Frankenstein – *Frankenstein* by Mary Shelley (1818)
- Dr Jekyll – *The Strange Case of Dr. Jekyll and Mr. Hyde* by Robert Louis Stevenson (1886)
- Dr Mabuse – *Dr Mabuse der Spieler* by Norbert Jacques (1922)
- Dr Moreau – *The Island of Dr. Moreau* by H.G. Wells (1896)
- Dr No – *Dr. No* by Ian Fleming (1958)
- Dr Octopus (Otto Octavius) – *Amazing Spiderman* issue #3 (Marvel Comics, 1963)
- Dr Psycho (Edward Cizko) – *Wonder Woman* issue #5 (DC Comics, 1942)
- Dr C.A. Rotwang – *Metropolis* (1927)
- Dr Strangelove – *Dr. Strangelove or: How I Learned to Stop Worrying and Love the Bomb* (1964)

the *Diagnostic and Statistical Manual of Mental Disorders* (psychiatry's catalogue of officially recognised conditions) listed Dissociative Identity Disorder as a recognised and treatable illness.

Genetic engineering

In H.G. Wells' novel *The Island of Dr Moreau* (1896), a rampant, drooling vivisectionist is secretly conducting surgical experiments with the goal of transforming animals into humans. Though the aim was to create a race without malice, the result of the doctor's insanity is a race of half-human, half-animal creatures that lurk in the island's jungles, only marginally under Moreau's command.

Wells' book was the most notable example of a handful of early stories that featured the deliberate 'engineering' of living creatures. It was written at a time when the scientific community was engaged in an impassioned debate on animal vivisection. Indeed, pressure groups were even created to confront the issue: the British Union for the Abolition of Vivisection was formed in 1898.

By 1924, little more was known of the biochemistry of genetics. Even so, British biologist J.B.S. Haldane foresaw our genetic future. His remarkably prophetic *Daedalus, or Science and the Future* (1924) divined a day when scientists would engineer a solution to the world's food problem, and modified children, born from artificial wombs, would represent a genetic improvement of humankind.

But Haldane was also a keen and shrewd populariser of science. He realised that there would be an acute reaction against the 'blasphemous perversions' of direct genetic manipulation. He was not to be disappointed.

Haldane was a friend of the Huxleys. Ideas from Haldane's optimistic *Daedalus*, such as ectogenesis (the growth of foetuses in simulated wombs), had greatly influenced Aldous' *Brave New World* (1932), in which ectogenetic

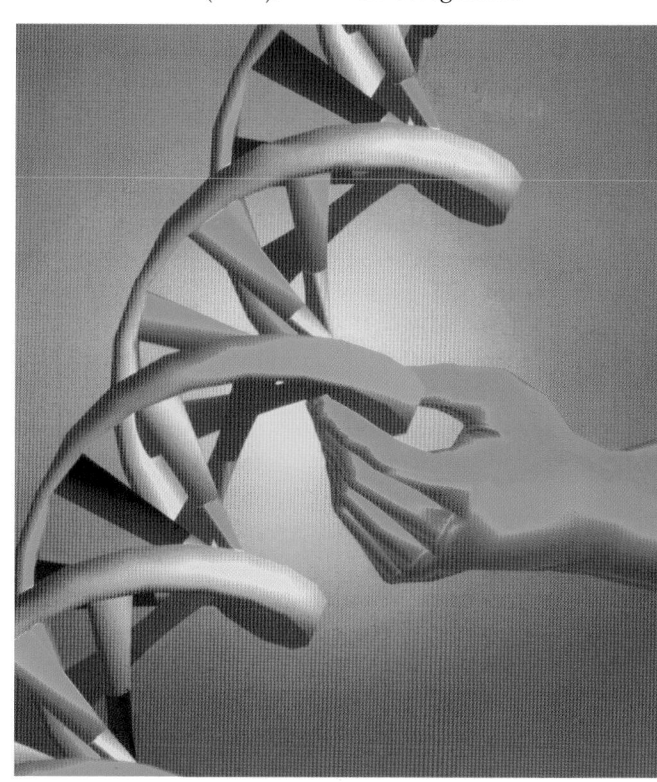

embryos are engineered to fit them for life as 'alphas', 'betas' or 'gammas'. Huxley's extrapolation of a future in which there is no war, no poverty and no pain through the application of genetics harboured dark secrets. A future with diminished genetic variance strips us of our humanity. Julian Huxley, friend to Wells and Haldane, wrote a notable story along the same lines in *The Tissue-Culture King* (1927).

By the 1950s the code was discovered and the structure of DNA understood. Since then the genetic engineering of bacteria has become commonplace. But Haldane's prediction of public mistrust persists. Despite science fiction's innate technophilia, it has shown little support for genetic engineering.

A new wave of fiction predictably surfaced after the 1960s. *Doomwatch* (1970–72), a BBC television series about an agency dedicated to preserving the world from dangers of unprincipled scientific research, reflected the anxiety with which biological engineering was regarded.

As genetic research makes rapid progress, authors have acquired a better sense of what actually goes on in real labs. Michael Crichton's *Next* (2006) is a techno-thriller about our bio-technological world. Throughout the novel, Crichton explores a world dominated by genetic research, corporate greed and legal conflict. *Next* features governments and private investors who spend billions of dollars each year on genetic research. It follows a genetic researcher as he produces a transgenic ape with some human features and the psyche of a young child. His family struggles to raise the chimera, as they attempt to hide the true nature of the ape's genetic makeup. And a leading genetic research company is embroiled in a lawsuit with a cancer survivor whose cells it has taken without his knowledge. The company also develops a 'maturity' gene that seems to transform social deviants into sober, responsible individuals.

Is this our future? A frighteningly bizarre world of gene-mongering scientists and biotech profiteers leading us into a strange moral wilderness? Only time will tell. Meanwhile, it seems many writers reflect, only too willingly, public anxieties over Haldane's 'blasphemous perversions' of genetic engineering.

LONGEVITY

The elixir of life. The fountain of youth. Writers have long dreamt of cheating time, maybe even cheating death.

Modern science fiction movies, such as the screen version of graphic novel *The League of Extraordinary Gentlemen* (2003), often lean heavily on myth. The film features one Dorian Gray, a

- Dorian Gray: what he really looks like.
- Dorian Gray: what we see.
- Godzilla slugs it out with Hedorah.

man who does not age, based on a character from Oscar Wilde's only novel. It is a painted portrait of the narcissistic Gray that ages, and not the man himself. But the deal is to his detriment, as his actions finally impact upon his soul. The film takes the novel's ideas still further. The portrait also prevents Gray from suffering injuries.

As the genetic code has unravelled, research in biotechnology has sparked much speculation that the technologies of longevity are now a real prospect. Some researchers believe the key to aging may be found in telomeres. These are tiny DNA-based structures at the end of chromosomes. Simply put, the telomeres shorten with each cell division. When the telomeres are gone, the cell dies. So, if we can lengthen the life span of cells, we will live longer in the future.

OIL-EATING MONSTERS

It was the patently ridiculous Japanese monster flick of 1968 *Yongara* that introduced us to the idea of an oil-eating monster. In the film the mutated 100-foot sea snail was defeated by an eight-year-old boy and an industrial sized tank of itching powder. Next we have Hedorah from the 1971 film *Godzilla vs. Hedorah,* which consists of billions of microscopic creatures that form a living black sludge capable of consuming oil and other petrochemicals. Fast forward to 1998 and the latest incarnation of the lifeform in the animated children's show *Godzilla: The Series* (spun from the 1998 film re-make of the monster classic,) and another oil-eating creature is discovered. Only this time the villain is called PEMC (Petroleum Eating Microbe Colony).

Whatever the name, the threat remained the same. Mankind, with its reliance on petrochemicals, was in peril. As we slip into the twenty-first century we find that 2007 witnessed two noteworthy events: the release of the R. Scott Reiss novel *Black Monday,* which features a world devastated by a microbe which contaminates its oil supplies, and the announcement by American environmental scientists who discovered bacterial organisms that live in and 'eat' heavy oil and natural asphalt at the Rancho La Brea tar pits of Los Angeles, California. Time to break out the pushbike.

TEST-TUBE BABIES

We'd seen them coming in *Brave New World* (1932), Aldous Huxley's disconcerting picture of twenty-sixth century London. In the book, Huxley foresees a future of biological engineering and reproductive technology in which multiple births are gestated in a lab and then 'decanted'.

Huxley was among the first to realise that

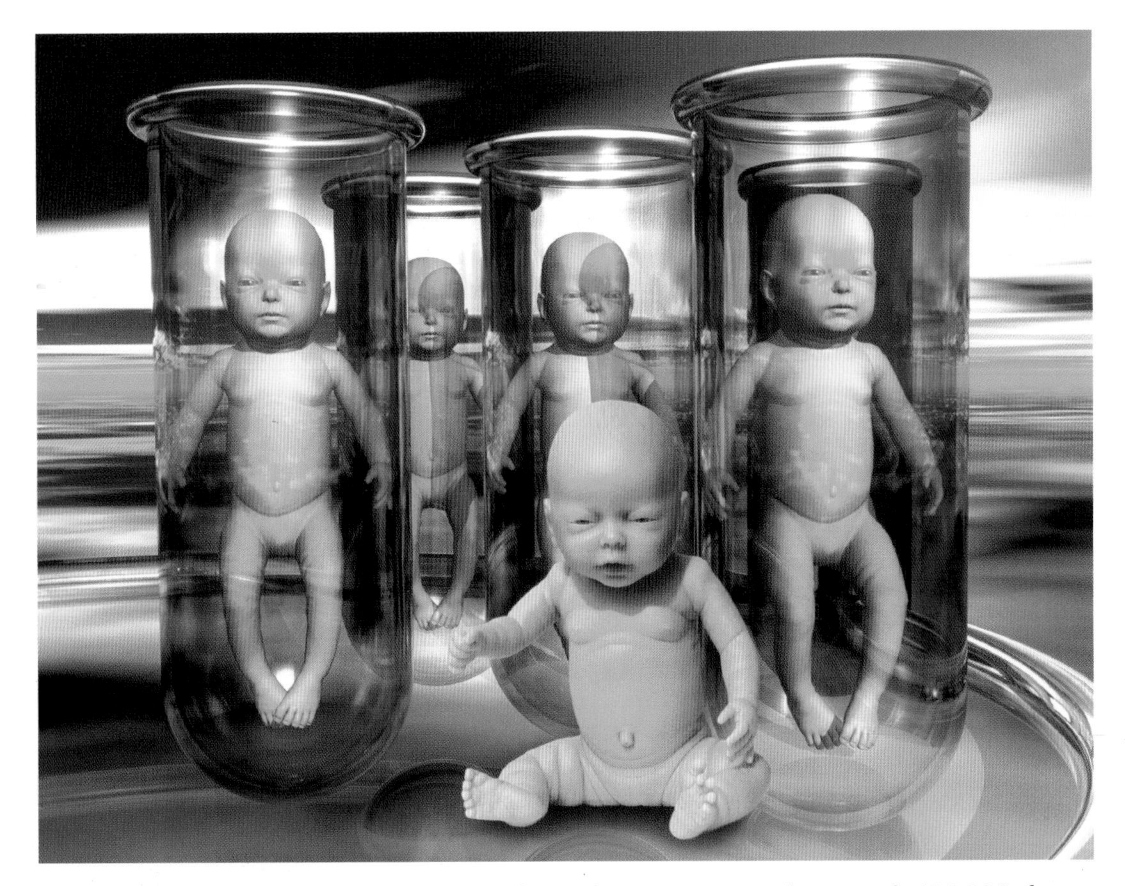

reproductive technology would lead to bioethical issues. More recent science fiction works such as *Gattaca* (1997) and *The Island* (2005) also focus on the impact of reprogenetic technology. Such techniques may have the latent potential to alter the assumptions that lie behind existing systems of sexual and reproductive morality.

The world's first factual test-tube baby was Louise Brown, born on 25 July 1978 and conceived through the then-new method of *in vitro fertilisation* (IVF). Since Louise, ART (Assisted Reproductive Technologies) has been used throughout the developed world. In the United States, 70,000 children have been born through ART, with 45,000 through IVF.

According to a 2003 report by the RAND Corporation (a non-profit research institute), there are currently around 400,000 frozen embryos stored in fertility clinics in the US alone. Each of these potential humans awaits a decision. Either the parent(s) will decide upon some date of future delivery, or the embryo will be kept in cryogenic suspended animation indefinitely.

MONSTERS FROM THE ID

Id, ego and superego. That's how Sigmund Freud saw the human mind in his psychoanalytic model of the early twentieth century. In Freud's structural model, each of these three component parts vie for supremacy, and the outcome of the ongoing conflict is reflected in human behaviour.

Plenty of science fiction stories feature this hidden self. Perhaps the most famous is the excellent 1956 movie *The Forbidden Planet*.

↑ In the US 70,000 children have been born through Assisted Reproductive Technologies.
→ 'Monsters from the id!'

The film features a spaceship crew exploring an alien world threatened by a monster. The indigenous population has died out, but two humans survived from an earlier mission, including a Dr Morbius.

An archetype of cold intellect, Dr Morbius harbours powerful passions that never reach the surface. The beast that repeatedly attacks the spaceship crew turns out to be Morbius' own savage unconscious, or 'Monsters from the id!' as one crew member cries.

The movie's message is plain: beneath the rational surface of science, made flesh by Morbius, lurks a demon. And the more we multiply our power through technology, the more we feed our demons.

Though Freud's theory has to some degree fallen out of scientific favour, it remains to this day a cultural shibboleth, thanks to movies such as *The Forbidden Planet*.

BLACK HOLES (EAT THE EARTH)

The idea of an object so dense that it emits no light was first mentioned by the geologist John Michell in a letter written to Henry Cavendish in 1783. However, it was not until 1915 when Karl Schwarzschild extrapolated from Einstein's Theory of General Relativity that a robust model for black holes was put forward. From that point on science fiction started to speculate about artificially created black holes that could create havoc. The 1941 short story *The Vortex Blaster* by E.E. 'Doc' Smith features a hero who goes around snuffing out such destructive phenomena, whilst the 2000 science fiction television series *Andromeda* features a race of marauding **ALIENS** who fire artificially created mini black holes at planets as a weapon.

It is just such mini black holes that could be created as the Large Hadron Collider based at CERN (European Organization for Nuclear Research) in Switzerland is switched on. Some scientists fear the worst from this high energy particle accelerator if predictions of Superstring Theory prove accurate. However, it is also theorised that small black holes could evaporate quickly due to the existence of the controversial phenomenon known as Hawking Radiation. According to experimental physicist Greg Landsberg of Brown University, the chance of planetary annihilation by manufactured black holes 'is totally miniscule'. Very reassuring.

SPARES

Mary Shelley's *Frankenstein* (1818) is the tale of Victor Frankenstein's alien creature. Victor intends his creature to be beautiful, and builds a mechanically sound but grotesque man using cadaver spares from charnel-houses. Only when he rejects the dark arts of the old-world alchemists and turns to the new unbridled science does Victor succeed in his terrible triumph of creation.

Between the first edition of *Frankenstein* in 1818 and the second in 1831, the first volume of Charles Lyell's *Principles of Geology* (1830) was published, dramatically increasing the age of the Earth. The new geologists were industriously rooting reptilian bones out of the mud, bringing the extinct monsters back to life.

The idea of Frankenstein's creature leaps into the science fictional future of artificial life. It is the nature of life itself that is under the microscope, the quest to unravel the agency through which inanimate matter is given the vital spark of life.

Less than ten years after *Frankenstein* came the Burke and Hare murders. The nineteenth century saw science rapidly develop our understanding of human anatomy. But before 1832, there was a lack of cadavers for the study of anatomy in British medical schools, such as the one in Edinburgh. As medicine began to blossom, demand mushroomed.

The root of the supply problem was that the only legal supply of cadavers – the bodies of executed criminals – was falling due to a sharp reduction in the execution rate. This state of affairs attracted criminals willing to get cadavers by any means.

The Burke and Hare killings (also known as the West Port murders) were committed in Edinburgh between 1827 and 1828 by William Burke and William Hare. Understanding the economic principle of supply and demand, Burke and Hare made their victims insensible with drink, smothered them, and then sold the still-warm corpses of their 17 victims to the Edinburgh Medical College for dissection. Their principal customer was Professor Robert Knox. Understandably, the activities of body-snatchers (aka resurrectionists) gave rise to public fear and revulsion.

In 1967, the same year that Professor Christiaan Barnard made the first successful human-to-human heart transplant at Groote Schuur Hospital, Capetown, South Africa, author Larry Niven invented the fictional crime of organlegging in *The Jigsaw Man*. A portmanteau of the words 'organ' and 'bootlegging', and meaning the piracy and smuggling of organs, Niven's vision was of a future where the transplant of any organ was medically possible.

In theory, organ banks could be used to extend life indefinitely. In Niven's 'reality', this proved tricky. To maintain communal organ banks, donors are needed. But when the death rate is reduced (via the organ banks), the number of donors decreases; another problem of supply and demand. The system is fundamentally flawed.

In the real world, organ theft is the stuff of urban legend. However, BBC reports in 2006 suggested that the sale of organs taken from executed death row inmates appears to be thriving in China. One associated hospital said it could provide a liver at a cost of £50,000.

Michael Marshall Smith's *Spares* (1996) foresees a future where 'farms' of clones

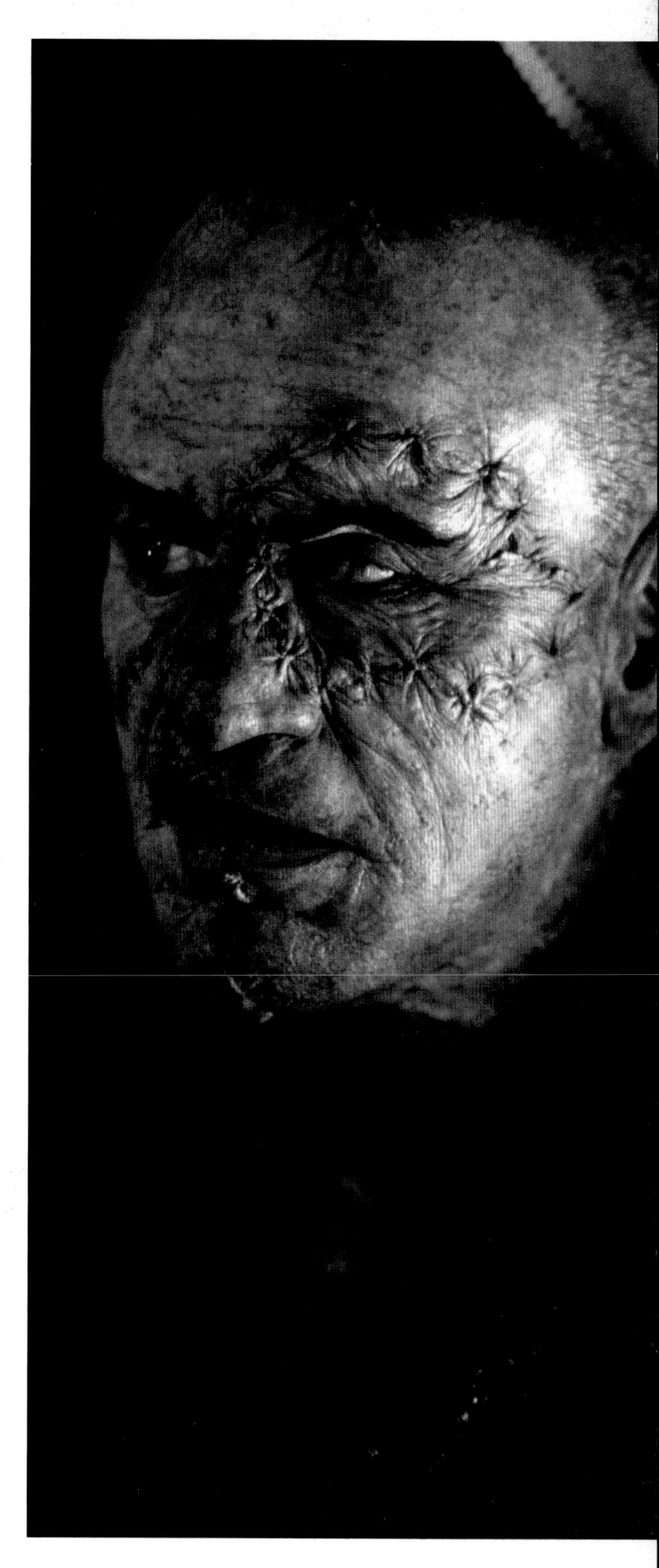

Testing the superconducting magnets that form part of the Large Hadron Collider particle accelerator at CERN.

Professor Greg Landsberg tries to look reassuring.

Are you looking at him? Robert de Niro as the Creature in *Mary Shelley's Frankenstein* (1994).

(spares) are kept as the ultimate insurance policy of the rich and powerful. Lose an eye, limb or vital organ, no problem. Money talks. Your body double is mutilated, and you get your replacement part.

Smith's book bears an uncanny resemblance to Michael Bay's movie *The Island* (2005). It is 2019. Most of the outside world has been contaminated. A community of people, rescued from the toxic environment, believe they are living in a utopian, isolated colony. The reality is darker still. The colonists are actually clones, walking, talking spares, whose sole purpose is to provide medical insurance for their celebrity sponsors.

MODSTARS

'Like a puppet on a string' sang Sandy Shaw in her 1967 Eurovision Song Contest winning entry. And now you have the chance for real, by becoming a 'modstar'. In the 1920 silent film classic *The Cabinet of Dr. Caligari*, the eponymous madman controls his somnambulist Cesare. By 1984 William Gibson had updated the science in his cyberpunk classic *Neuromancer*. There he featured 'meatpuppets', individuals whose actions were controlled by someone else via a wireless link. They had no choice and women frequently found that they had been coerced into sexual indiscretions as a result, renting themselves out for profit.

Now a group of entrepreneurs in New York City have set up the *Mod My Life* website through which they offer users the chance to log on and control the actions of a designated individual. There are no strings, at least not physical ones, for they employ wireless video cameras and a remote internet connection. The 'modstar' is controlled not by one person, however, but by all the registered users who vote on his or her actions. Presumably, the modstar is free to balk at illegal or dangerous activity, unlike the meatpuppets of Gibson's book. However, the principle is the same. Are you happy to become a mindless somnambulist unaccountable for your actions? Do you want to be a modstar?

HUMAN CLONING

History is replete with the idea of copies; simulacrums of ourselves that we encounter as if we were looking in the mirror. One ancient idea is that of the *Doppelgänger*, a spirit which takes our exact form and is usually a herald of death. Mary Shelley, the author of *Frankenstein*, wrote that her husband, the poet Percy Shelley, had seen his *Doppelgänger* two weeks before drowning.

Cesare in action from *The Cabinet of Dr. Caligari* (1920).
Today's modstar in action.
Perfect copy: monozygotic twins (when the fertilised egg splits in two) are genetically identical.

Despite these mythic explorations, science tells us that clones have existed for a long time. Monozygotic clones or identical twins are formed when a single fertilised egg divides into two separate embryos. Whilst genetically they are completely identical (sharing the same DNA), their appearances and their characters are not. The idea of twins has historically been employed many times in stories dating as far back as the founders of Rome, Romulus and Remus, in 770BC. Romulus slew his brother following a disagreement, an action which could well have contributed to the fascination with the concept of the 'evil twin'. This proved fertile ground for film with both early serials like

the 1937 version of *Dick Tracy*, the 1939 version of *The Man in the Iron mask* and Charlie Chaplin's 1940 anti-Nazi *The Great Dictator*, where Chaplin plays a poor Jewish barber who is a dead-ringer for Adenoid Hynckel, a savage and hilarious caricature of Hitler.

Science fiction explored the deliberate attempts to clone humans with arguably the first sustained and explicit story coming from A. E. Van Vogt. In his 1945 space opera novel *The World of Null-A*, the hero Gilbert Gosseyn dies but then lives on with his consciousness intact in a series of cloned bodies, all the while being manipulated by a mysterious force.

Slightly less fantastic was the 1971 film *The Resurrection of Zachary Wheeler*, starring science fiction luminary Leslie Nielson. This thriller tells the tale of a secret committee which has instituted a covert medical program designed to create back-up clones of important world leaders so that they will have material for organ transplants if accidents occur. The plot revolves around a reporter who threatens to uncover the secret. With discussions on cloning, stem cells and the ethics of DNA technology, this little known film was prescient in its discussions at a time when the science was a long way off.

Another attempt to explore this type of plot came in 1978 when the film adaptation of Ira Levin's book *The Boys from Brazil* mixed in South American Nazi exiles to create the threat of a fourth Reich. This was to help create worldwide interest in cloning at the time along with the book *In His Image: the Cloning of a Man* (also 1978) by the science journalist and *Time Magazine* medical correspondent David Rorvik. In the book, Rorvik claimed to have been part of a successful attempt at deliberate human cloning. Since largely discredited (alongside later 2002 claims by the religious group The Raelians that they had cloned a little girl they named 'Eve'), it was not until 2004 that science felt it had made a robust scientific breakthrough.

South Korean geneticist Hwang Woo-Suk of Seoul National University claimed to have built on the breakthrough in **ANIMAL CLONING** that was Dolly the sheep to produce 30 cloned human embryos to the one-week stage. Published in the reputable journal *Nature*, it was only in 2005 that this (and later work) was revealed as fraudulent. It seems we aren't in any danger of meeting our *Doppelgängers* just yet.

ARTIFICIAL BACTERIA

Imagine a future in which modern warfare was obsolete. A utopia? Not likely, instead the weapon of choice is artificial bacteria designed to specifically target certain elements of the population. Sound far fetched?

Geneticist Hwang Woo-Suk gets unwelcome coverage from the world's media.
Microscopic weapons of mass destruction could be created in the lab.
Water: now with added benefits.

Well it probably was when Raymond Z. Gallun wrote his 1937 novel *Seeds of Dusk*. There, the hero becomes the inadvertent carrier of a tailored bacterium that is threatening to destroy mankind. A similar threat is extended to just over half the world's population, the women, in science fiction favourite Frank Herbert's 1982 book *The White Plague*.

However, it was not until 2007 that this took a major step closer to being a reality when the patent for the world's first artificial bacterium was applied for. A team assembled by Craig Venter, one of the leading lights behind mapping the Human Genome (which successfully sequenced human DNA) applied for the patent of *Mycoplasma laboratorium*. This artificial product was created by stripping away one hundred and one 'excess' genes from the bacterium *Mycoplasma genitalium*, which causes urinary tract infections. This left the researchers with the basic 'chassis' of the bacterium. They say that using this frame they can build new bacteria that will accomplish a variety of different tasks, from speeding up the production of ethanol and hydrogen for fuel to more effectively trapping carbon dioxide as a way of slowing down climate change.

PHARMACOLOGIES

Imagine a world in which you were forced to take drugs, a world in which the average person in the street had little if no choice about adding something to their system that wasn't naturally meant to be there. Science fiction has had a long flirtation with drugs, from the elixirs of Dr Jekyll, Dr Moreau and the Invisible Man to the smart drugs of cyberpunk, which helped man and machine combine. However, it is Aldous Huxley's 1934 classic *Brave New World* which dominates the landscape. His vision of a regimented, mechanised and drug-induced utopian society arranged for optimum social stability serves as a stark warning about the widespread consumption of pharmacology. His fictional society practised compulsory mass medication. Surely that's something we would never do.

Or would we? There are large portions of the world where medical science is at odds with itself because of just this issue. The controversy persists around the safety and propriety of water fluoridation – the practice of adding fluoride to drinking water to improve dental health. Countries and scientists across the world disagree. America, Canada and Britain allow it. France, Germany and Japan do not. Whilst the health benefits and hazards are discussed, the issue of consent was raised as a problem by Huxley long before the real world caught up.

THE INVISIBLE MAN

The perennial dream of teenage boys, the idea of being invisible has haunted myth, magic and the imagination for millennia. In many legends, angels and demons were often invisible, or could become so at will.

In ancient and medieval astronomy, the crystal spheres that held up the Sun, Moon and planets were also invisible.

The modern day infatuation began with Wells. His famous novel *The Invisible Man* (1897) spawned James Whale's influential 1933 movie adaptation and countless comic strips and TV series.

Wells daubed the myth of invisibility with a scientific veneer. Wells' Invisible Man is a scientist. His theory is this: if a person's refractive index is changed to precisely that of air, and his body does not absorb or reflect light, then he will become invisible. This he does. But he cannot become visible again, which drives him mad.

To avoid such insanity, physicists have donned invisibility cloaks, a form of **OPTICAL CAMOUFLAGE**, rather than swallow a chemical formula. One such cloak, developed by Professor Vladimir Shalaev at Purdue University in Indiana, is made of angled tiny metal needles that force light to pass around the cloak. The wearer appears to vanish, without the Wellsian drawback of lunacy.

A working prototype is expected by 2010. At the moment though, there is one major drawback: the current design can only bend the light of a single wavelength at a time, and does not work with the entire frequency range of the visible spectrum.

FACE TRANSPLANTS

The idea of the face transplant to restore the severely damaged or badly scarred had its first mention in the October 1943 edition of *Detective Comics* entitled 'The End of Two-Face'. There the notorious villain Harvey 'Two-Face' Kent, a former district attorney scarred by acid, is rehabilitated by Batman and receives surgery before being re-united with his fiancée, Gilda. In film, the 1960 movie *Les Yeux Sans Visage* (Eyes Without a Face) features a mad scientist who forcibly transplants young girls' faces onto his disfigured daughter, whereas in the 1964 Japanese *Tanin No Kao* (The Face of Another) it is the scientist himself who gets a new face.

In 1994, Indian surgeons performed the world's first full face transplant on a nine year-old girl who had been horribly disfigured by a threshing machine. Technically, it was a face *replant* as the recipient of the face was also its original owner.

In 2005, Isabelle Dinoire underwent a fifteen-hour operation in Amiens, France, to receive a new nose, mouth and chin from a brain-dead donor after her own face was mutilated in an attack by her pet dog.

Claude Rains' *Invisible Man* can't see the point of it all.

The 1933 film was advertised with the tagline 'Catch me if you can!'.

A restored Isabelle Dinoire.

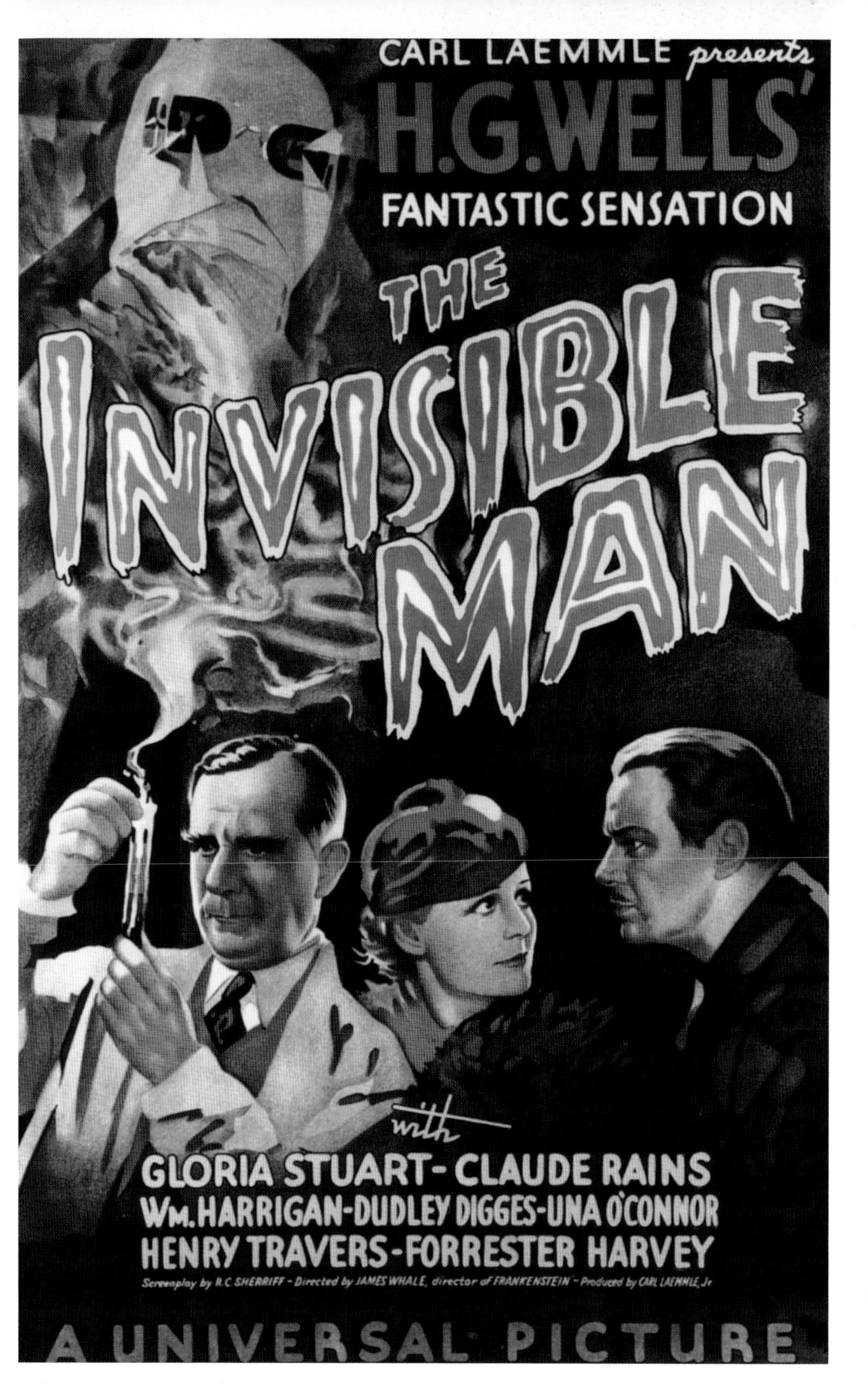
CARL LAEMMLE presents
H.G.WELLS'
FANTASTIC SENSATION
THE
INVISIBLE
MAN
with
GLORIA STUART-CLAUDE RAINS
Wm. HARRIGAN-DUDLEY DIGGES-UNA O'CONNOR
HENRY TRAVERS-FORRESTER HARVEY
Screenplay by R.C. SHERRIFF - Directed by JAMES WHALE, director of 'FRANKENSTEIN' - Produced by CARL LAEMMLE, Jr
A UNIVERSAL PICTURE

MICROBIAL MONSTERS

The world might end not with a bang but with a sneeze. Forget the **ATOMIC BOMB**; it's your friend or neighbour who could prove more deadly if they are carrying an infectious disease. Scientists at the World Health Organisation (WHO) monitor the health of the population of the world. They classify our susceptibility to widespread outbreaks, known as pandemics, on a six point scale. In 2007 the world stood at point three.

The WHO believes that humanity is now nearer to another major casualty-causing pandemic than at any time since 1968 when the last of the previous century's three pandemics occurred. What is the threat from? It could be HIV (the virus that causes AIDS) which the WHO has classified as a worldwide pandemic. However, transmission vectors (the means by which the disease is passed from host to host) mean its spread and onset are gradual.

It could be Severe Acute Respiratory Syndrome (SARS) which caused a smaller pandemic between 2002 and 2003. However, the current major candidate is the H5N1 strain of the influenza virus, often referred to as 'bird flu'. For some years scientists have been warning both governments and the public of the consequences of a major outbreak. To put it in perspective, the H1N1 influenza outbreak of 1918–19 killed between twenty-five and fifty million people worldwide, and that was before easy international travel made the world a smaller place. Yet for all the dire prognostications of scientists, the apocalyptic visions of a globe wracked by disease have seemed to fall on deaf ears. That's where science fiction comes in.

Although disease has been with humanity since the very beginning, it is science fiction that has been able to capture the public's

↑ Coughs and sneezes spread diseases: in 2007 the World Health Organisation estimated our susceptibility to a pandemic stood at point three on a six-point scale.

↗ Let's hope any pandemic isn't as messy as *Resident Evil* (2002).

imagination as to the societal consequences that a global pandemic could cause. From Mary Shelley's 1826 second novel *The Last Man* in which a sole survivor of a globe depopulated by disease wanders amidst the debris, science fiction has helped us to come to terms with one of our possible futures. Shelley's work was hugely influential on H.G. Wells', who in his famous 1898 work *The War of the Worlds* had microbial monsters assist us in defeating the invading Martians. Of course it's entirely possible that things could be reversed. The astronomical theory of Panspermia advanced by Fred Hoyle and Chandra Wickramasinghe in the 1960s argues that microbes could come from space to Earth, a scenario explored by the 1969 Michael Crichton book *The Andromeda Strain* and more recently by the 2005 television series *Threshold*.

However, it is far more likely that such outbreaks will be terrestrial in origin. Works like Jack London's 1912 *The Scarlet Plague*, which resonated during the 1918–19 H1N1 'Spanish' influenza outbreak, have already demonstrated that the public often appeals not to logic and science but to the scenarios conjured up in the fevered imaginations of fiction. The recent 2005 film *28 Days Later* took great pains to show a post-apocalyptic vision of a London nearly emptied by a virus outbreak. This was followed by a no less serious attempt to show a London in peril in the 2007 film *Children of Men*, in which a virus has caused mass sterility in the population and threatens the future of humanity. Even recent computer games have sought to explore this motif within the public consciousness with games like *Resident Evil* (1996) and its sequels (subsequently made into films). These games have introduced an additional factor, however – man's own negligence, with the diseases having escaped from laboratories where they were being developed by commercial concerns. These **ARTIFICIAL BACTERIA** are depicted as very real monsters which force the player to fight their way past friends and family. So if you want to save the world, use a tissue.

SUPERSERUM

The hero (or monster) within us is but a sip away.

With a gulp of a strange concoction, Dr Jekyll was able to awaken Mr Hyde. In the closing decades of the nineteenth century, writers were beginning to explore the influences of science on the human frame. And what did this post-Darwinian scientist Jekyll find once he'd necked his potion? An evil and lustful nature, something gothic at the heart of the human.

And so began another fictional obsession: heroes and villains who possess special powers or abilities, gifted by the ingestion of a specialised compound.

Captain America is one of a kind in the *Marvel Universe*. But only because the inventor of a formula, Super-Soldier Serum, got shot after administering the good Captain.

Consumers seem to have become similarly obsessed with superfoods, types of food believed to have life-transforming health benefits such as helping us to stay younger-looking, live for longer and improve our mental capabilities. Blueberries are considered a superfruit due to their high concentration of antioxidants and dietary fibre.

But the most potent superfood is cacao ('ka-cow'), the raw, un-cooked form of chocolate. Scientific investigation has shown that raw cacao contains over 100 chemical constituents, including amino acids, vitamins, minerals, polyphenols, alkaloids, phospholipids, serotonin, tryptophan and protein, to name but a few.

ARTIFICIAL ORGANS

It seems that we are entering the age of the **CYBORG** when man-machine hybrids are the answer to our health problems. Artificial hearts, joints and even aids to erectile dysfunction all combine electronic wizardry with human ingenuity to provide a solution. But what if there was another way? What if we could just grow new organs instead?

This was Larry Niven's vision in his 1968 book *A Gift From Earth*. In a grim foretelling of what is allegedly happening in some of China's prisons at the moment, the book depicts prisoners in a colony in space having their organs harvested for public consumption. This abhorrent system is only broken when a cargo of artificially grown organs arrives from Earth to break the monopoly. In January 1999 the first laboratory grown organ, a bladder, was implanted in a mammal (in this case a dog). It took only seven more years before the first successful human implants of artificially grown bladders were implanted into seven young people aged between four and nineteen. Samples of the patients' own bladders were grown into full size organs by doctors who then implanted them back into the host. Since that time, livers have been added to the list with other organs scheduled for implantation soon.

An artificial bladder grown from cultured bladder cells in just four weeks.

Mind over matter: Famke Janssen is telekenetic Jean Grey in *X-Men: The Last Stand* (2006).

PSYCHIC POWERS

Folklorist and occultist notions of power have been with us for centuries.

Had Isaac Newton not been inspired by the occult concept of action at a distance, he might not have developed his theory of gravity. Newton's use of the occult forces of attraction and repulsion between particles led British economist John Maynard Keynes, who in 1936 acquired many of Newton's writings on alchemy, to suggest that, 'Newton was not the first of the age of reason: he was the last of the magicians'.

The study of the occult, usually meaning 'knowledge of the paranormal' in contrast to 'knowledge of the measurable' is associated with hidden wisdom. For the occultist such as Newton it is the study of 'truth' a deeper and more profound truth that lies beneath the surface.

Much science fiction has been written about this sense of a deeper spiritual reality that extends beyond pure reason and the physical sciences. And for many writers, rediscovered powers based upon such a hidden reality might be developed in the course of man's future evolution.

Psi powers is the name given to the full spectrum of mental powers, which are an assumed element of this hidden reality. The name stems from the study of the pseudoscience of parapsychology, and is a widely used term in the science fiction tradition. The term was particularly prominent during the 'psi boom' that John W. Campbell Jr promoted in *Astounding Science Fiction* magazine during the early 1950s.

Indeed, a related term, psionics – a term derived from combining the psi, signifying parapsychology, with electronics – arose in the late 1940s and early 1950s. Psionics revolved around the application of electronics to psychical research.

An early psionics instrument was the Hieronymous Machine. Ostensibly the invention of Dr Thomas Galen Hieronymous, but promoted widely by Campbell in *Astounding Science Fiction* editorials, Hieronymous Machines were mock-ups of real machines. They allegedly worked by analogy or symbolism, and were directed by psi powers. For example, one could create a receiver or similar device of prisms and vacuum tubes, but instead use cardboard or even schematic representations. Through the use of psi powers, such a machine would function as would the 'real' equivalent. Campbell claimed that these machines actually did perform this way. Unsurprisingly, the concept was never taken seriously elsewhere.

Still, science fiction writers speculated on a future where man would harness such mental capabilities. Typical is Arthur C. Clarke's *Childhood's End* (1953). The dawning of a space age is suddenly aborted when enormous alien spaceships appear one day above all of the Earth's major cities. The **ALIENS**, the Overlords, quickly end the arms race and colonialism.

After one hundred years, human children start displaying psi powers. They develop telepathy and telekinesis. They become distant from their parents. The Overlords' purpose on Earth is finally revealed. They are in service to the Overmind, an amorphous extraterrestrial being of pure energy. The Overlords are charged with the duty of fostering humanity's transition to a higher plane of existence and merger with the Overmind.

Interestingly, in the preface of a 1990 reprint and partial re-write of *Childhood's End* Clarke attempted to unravel pseudoscience from his extraterrestrial message: 'I would be greatly distressed if this book contributed still further to the seduction of the gullible, now cynically exploited by all the media. Bookstores, news-stands and airwaves are all polluted with mind-rotting bilge about **UFOs**, psychic powers, astrology, pyramid energies.'

But Clarke believed in a future where man will nonetheless meet his superiors in space: 'the idea that we are the only intelligent creatures in a cosmos of a hundred billion galaxies is so preposterous that there are very few astronomers today who would take it seriously.'

TRANSCENDENCE

Einstein remarked, 'Reality is merely an illusion, albeit a very persistent one'. And mankind has long sought to transcend beyond our reality into a new existence. Although most frequently a religious theme, the idea that science can come to help us move beyond our physical limitations has been explored by science fiction in preparation for the technological innovations that some think are only round the corner.

12 cool science terms no-one understands

- The Big Bang
- Dark Matter
- Dwarf Star
- Googolplex
- Neutrino
- Quark
- Quasar
- Red Shift
- Starquake
- Supernova
- Syzygy
- WIMP (Weakly Interacting Massive Particles)

Starting with one of the first true pieces of science fiction, Mary Shelley's 1818 classic *Frankenstein*, writers have sought to pose difficult questions. Questions like what is beyond the death of the body? Can we evolve beyond our physical form? Could we live forever in some way?

In Edgar Allen Poe's only novel, *The Narrative of A. Gordon Pym* (1838), the captive 'savage' nu-nu is mystically transformed at the end of the book with no scientific explanation. Poe's work influenced H.P. Lovecraft whose early twentieth century visions of transcendence came to rely on the intervention of largely incomprehensible **ALIENS**. This was a theme picked up by Stanley Kubrick and Arthur C. Clarke in the influential 1968 science fiction film *2001: A Space Odyssey*. This science fiction classic is a three-part exploration of Nietzsche's ideas of transcendence from 'Apeman' to 'Human' to 'Starchild', with the creators of the mysterious monoliths helping at each stage. We are treated to a knowing nod in this direction by the inclusion of Strauss' *Also Sprach Zarathustra* (itself a homage to Nietzsche) as its iconic soundtrack. A similar theme had already been explored by Clarke in his 1953 work *Childhood's End*. In this novel, benevolent aliens help mankind to

➋ The transcendent Star Child from *2001: A Space Odyssey* (1968).

develop **PSYCHIC POWERS**. This leads to the evolution of mankind into a non-corporeal form, leaving our bodies behind.

A related exploration of non-corporeal existence was explored in the 1999 film *The Matrix* in which the hero Neo (an anagram of 'one') learns to control his simulated reality through the almost magical powers of his mind. He is able to halt bullets in mid-flight and bring the dead back to life. The conventional laws of physics just don't seem to apply.

Science is trying to catch up with Neo. In 2007 a major round of worldwide science funding, supported by the Templeton Foundation, was announced. The aim of this fund – Science and Transcendence Advanced Research Series (STARS) – is to pay for research by small groups of scientists and humanities scholars into the ways 'science, in light of philosophical and theological reflection, points towards the nature, character and meaning of ultimate reality.'

One of the groups to successfully bid for a grant is based at the Centre for Quantum Studies, George Mason University. They claim that 'our intuitive view of physical reality, a view that grows out of everyday experience, is often in tension with the world-view described by physics. It turns out that many of our most fundamental beliefs about the physical world lack a firm scientific foundation.' The team, led by Yakir Aharonov, Professor of Theoretical Physics, and Jeffrey Tollaksen, Professor, founder and director of the Centre for Quantum Studies, claim that in one example we experience time as having a certain direction: we are born, grow old, and die; eggs break, liquids mix, and our homes tend to get more disordered – not the other way round. Yet they point out that according to both classical and quantum theory, time has no such in-built direction. In the theory it doesn't matter if time runs forwards or backwards. The team are currently investigating the way that we view reality in order to understand how we can believe we have free will, even when physics says that the world evolves according to deterministic laws. The closer we get to understanding how the rules work, the easier it will be to learn ways that we can break them.

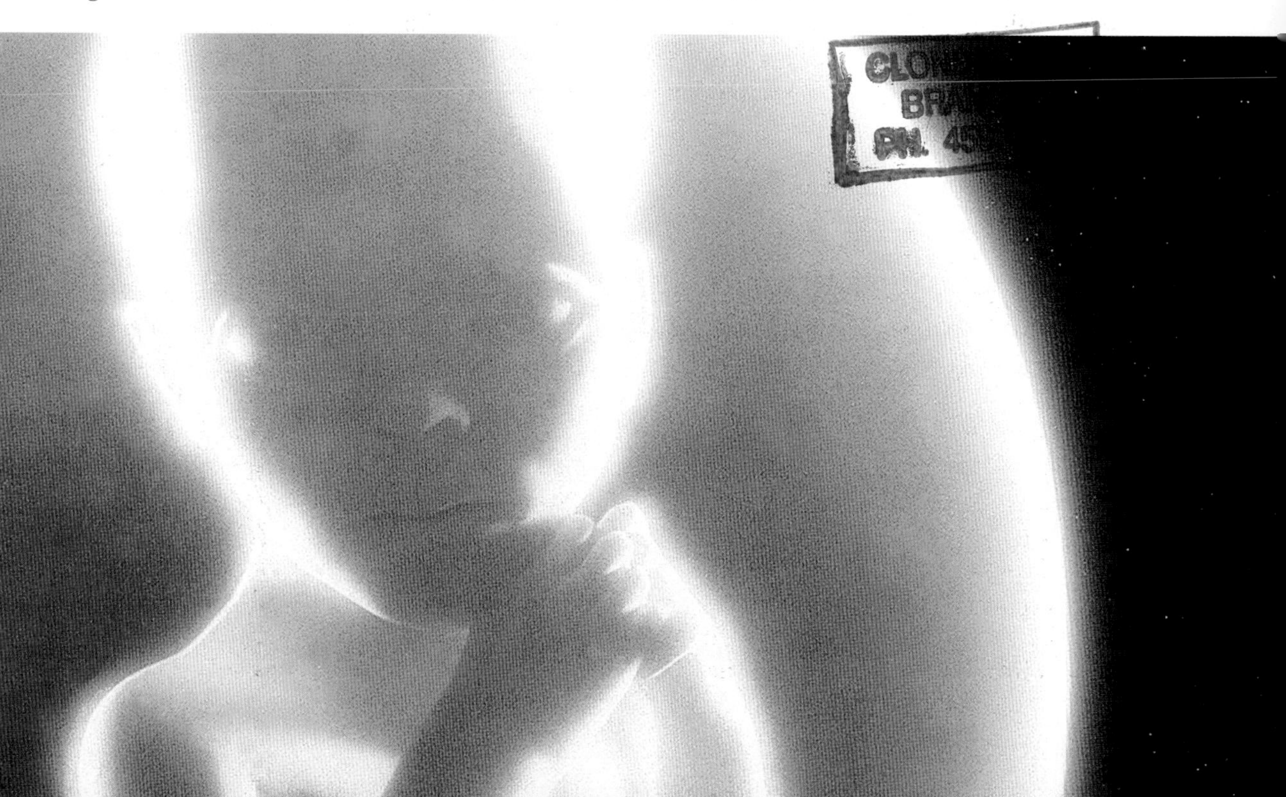

INDEX

PICTURE CREDITS

Pages 4–5, SSPL/NASA; 6t, SSPL/NASA; 6b, akg-images; 8, 20th Century Fox/The Kobal Collection/Hayes, Kerry; 9t, SSPL; 9b, Universal/Amblin/The Kobal Collection; 10t, akg-images/RIA Nowosti; 10b, Chris Butler/Science Photo Library; 11, Paramount Television/The Kobal Collection; 12t, © British Library Board. All Rights Reserved / The Bridgeman Art Library; 12b, SSPL/NASA; 13c, Michel Sarda/Soleri Archives; 13b, 2000 AD © and ™ Rebellion A/S; 14, Victor Habbick Visions/Science Photo Library; 15t, Bill Mueller/Vision Videogames LLC; 15b, akg-images/NASA; 16, Detleve Van Ravenswaay/Science Photo Library; 17t, Mary Evans Picture Library; 17b, SSPL; 18, NASA/Science Photo Library; 19, Detleve Van Ravenswaay/Science Photo Library; 20, possession of author; 21t, Omar Torres/AFP/Getty Images; 21b, Photo © Estate of Francis Bello/Science Photo Library; 22, Reprinted with permission from Microsoft Corporation; 23t, © Bruce Jensen; 23b, Ronald Grant Archive/ EON Productions; 24, © Oliver Reichenstein/informationarchitects.jp; 25, Ulf Andersen/Getty Images; 26, MGM/The Kobal Collection; 27, Chris Butler/Science Photo Library; 28, P.G. Adam, Publiphoto Diffusion/Science Photo Library; 29, 20th Century Fox/The Kobal Collection; 30, Dr. Jerry R. Ehman/NAAPO/www.bigear.org; 31, Mary Evans Picture Library; 32, Lucas Film/20th Century Fox/The Kobal Collection; 33, Alien III, sideview III (Work no.372) © 1979, H.R. Giger. All Rights Reserved. Courtesy of HRGigerMuseum.com; 35, Irwin Allen Prod./The Kobal Collection; 36, Science Photo Library; 37, Kudos Film and Television/BBC; 38t, Dreamworks/Warner Bros/The Kobal Collection/Cooper, Andrew; 38b, 20th Century Fox/Dreamworks/ The Kobal Collection; 39t, © Deborah Mallett; 39b, UFA/The Kobal Collection; 40, Scott Barbour/Getty Images; 41t, BBC Photo Library; 41c, © Kevin Perrott; 42, David A. Hardy/Science Photo Library; 44c, Emilio Segre Visual Archives/American Institute of Physics/Science Photo Library; 44b, SSPL/NMeM; 45, Touchstone/The Kobal Collection; 46, Paramount Television/The Kobal Collection; 47, 20th Century Fox/The Kobal Collection; 48, Giraudon/Bridgeman Art Library; 49, Gustoimages/Science Photo Library; 50, MGM/The Kobal Collection; 51t, Douglas McFadd/Getty Images; 51b, Lucas Film/20th Century Fox/The Kobal Collection; 52, Ethan Miller/Getty Images; 53, MGM/The Kobal Collection; 54, New Line/The Kobal Collection; 54–55, © Mart de Jong/Spacebox®/De Vijf bv Rotterdam NL; 55, Scott Olson/AFP/Getty Images; 56, Paramount/The Kobal Collection; 57, BBC Photo Library; 58, Tim Boyle/Getty Images; 59, © Vitaly S Alexius (svitart.com); 60, Warner Bros/The Kobal Collection/ Sorel, Peter; 61, Joe Tucciarone/Science Photo Library; 63, Maximilien Brice, CERN/Science Photo Library; 64, UFA/The Kobal Collection; 65, Peter Menzel/Science Photo Library; 66r, Moller International, USA; 66b, Gaumont/Ronald Grant Archive; 67t, Recon Optical; 67b, Steve Allen/Science Photo Library; 68, Lucasfilm/The Kobal Collection/Hamshere. Keith; 69t, Los Alamos National Laboratory/Science Photo Library; 69c, Orion/The Kobal Collection; 70, Carolco Pictures/Ronald Grant Archive; 71, UFA/The Kobal Collection; 72, Cordelia Molloy/Science Photo Library; 73, Columbia/The Kobal Collection; 74, Touchstone/ The Kobal Collection; 75, Shizuo Kambayashi/AP/PA Photos; 76, Mai/Mai/Time & Life Pictures/Getty Images; 77t, James King-Holmes/Science Photo Library; 77b, Warner Bros/The Kobal Collection/Boland, Jasin; 78t, Mike Nelson/AFP/Getty Images; 78c, Lorimar/Universal/The Kobal Collection; 79, Dirck Halstead/Time & Life Pictures/Getty Images; 80t, Claudio Edinger/ Liaison/Getty Images; 80b, Copyright 2008, Linden Research, Inc. All Rights Reserved; 81, Mary Evans Picture Library; 82, SSPL; 83t, Nikolaevich/Getty Images; 83c, SSPL/NMeM Daily Herald Archive; 84, 20th Century Fox/The Kobal Collection; 85, © Paul Glendell/Alamy; 86t, Mary Evans Picture Library; 86b, Melies/The Kobal Collection; 87, Shin Won-Gun/AFP/Getty Images; 88t, SSPL; 88b, Emilio Segre Visual Archives/American Institute of Physics/Science Photo Library; 89, Scott Camazine/Science Photo Library; 90–91, MGM/The Kobal Collection; 92, Universal TV/The Kobal Collection; 93t, Mark Smith/Science Photo Library; 93b, MGM/The Kobal Collection; 94, Win McNamee/Getty Images; 95t, Warner Bros TV/DC Comics/The Kobal Collection; 95b, Stephen Ferry/Liaison/Getty Images; 96t, Warner Bros/DC Comics/The Kobal Collection; 96b, Marvel/Sony Pictures/The Kobal Collection; 97t, © York Museums Trust (York Art Gallery)/Bridgeman Art Library; 97b, Science Photo Library; 98, Peter Newark Military Pictures/Bridgeman Art Library; 99t, BBC Picture Library; 99b, Peter Menzel/Science Photo Library; 100, MGM/The Kobal Collection; 101, Mehau Kulyk/Science Photo Library; 102c, MGM/The Kobal Collection; 102b, 20th Century Fox/The Kobal Collection; 103, The Kobal Collection; 104, Laguna Design/Science Photo Library; 105, Ronald Grant Archive; 106t, CERN/Science Photo Library; 106c, © Fermilab Visual Media Services; 107, Tri-Star/American Zoetrope/The Kobal Collection; 108t, Decla-Bioscop/The Kobal Collection; 108c, Mod My Life, Inc.; 109, Helen McArdle/Science Photo Library; 110t, Chung Sung-Jun/Getty Images; 110b, AJ Photo/Science Photo Library; 111, Garry Watson/Science Photo Library; 112t, Universal/The Kobal Collection; 113, Universal/The Kobal Collection; 112b, CHU Amiens via Getty Images; 114, Tim Vernon, LTH NHS Trust/Science Photo Library; 115, New Legacy/The Kobal Collection; 116, Sam Ogden/Science Photo Library; 117, 20th Century Fox/The Kobal Collection; 119, MGM/The Kobal Collection.

Every effort has been made to contact the copyright holders of images that appear in this book. Further information regarding copyright is welcomed.

ACKNOWLEDGEMENTS

I would like to thank my family, especially my wife Abigail and daughters Casey & Megan for putting up with me ignoring them for so long; my colleagues and students at The University of Glamorgan for their support, especially my co-author Mark; Damon and all at the Science Museum; and the guys who kept up my interest in all things science fictional and kept me sane including P.C, B.M, J.L, A.P, N.W, and especially J.C. *Neil Hook*

Thanks to my family for their patience through the long hours that writing this book consumed. Thanks to Professor Mark Rose for the economy and elegance of his anatomy of science fiction. *Mark Brake*

Thanks also to Jeremiah Solak at the Science Museum for getting the ball rolling...